IRRIGATION AND WEED MANAGEMENT LEARNING FOR BEGINNERS

IRRIGATION AND WEED MANAGEMENT LEARNING FOR BEGINNERS

Prof. Dr. K. Annadurai
Principal, Institute of Agriculture
Kumulur, Tamil Nadu

Prof. Dr.. P. M. Shanmugam
Institute of Agriculture, Kumulur,
Tamil Nadu

Dr. A. Vijayaprabhakar
TA *(Agronomy), Institute of Agriculture,*
Kumulur, Trichy, Tamil Nadu

SCIENTIFIC
PUBLISHERS

Published by
SCIENTIFIC PUBLISHERS (INDIA)
5 A, New Pali Road, P.O. Box 91
Jodhpur 342 001 (INDIA)
E-mail: info@scientificpub.com
Website: http://www.scientificpub.com

Print 2026

ISBN: 978-93-90749-43-0 (HB)
eISBN: 978-93-90749-59-1 (EBook)

Printed in India

TAMIL NADU AGRICULTURAL UNIVERSITY

Prof. N. Kumar, Ph.D., F.H.S.I. F.S.P.H.,
Vice-Chancellor

Coimbatone - 641003
Tamil Nadu, India

Foreword

Agriculture revolves around the principle of effective management of various inputs for crop production to meet out the human needs while maintaining or enhancing the quality of environment and conserving natural resources. Agriculture is the multidisciplinary science with the role of achieving more food production through day to day inventions by the scientists across the world. Agronomic principles particularly irrigation and wood management deal with employing scientific practices for improving the productivity of the crops for the benefit of human beings. For this, melding the young minds with intensive knowledge on water and weed management science and technologies is the need of the hour. The comprehensive information on irrigation and weed management for sustainable crop production is need of the hour.

The book titled "Irrigation and wood management learning for beginners" is written in a simple language to understand all the aspects of irrigation water and weed management in crop production. This book also focuses on understanding basic principles and practices in irrigation and weed science by the students of Undergraduate and Diploma level. Perhaps reading this book will kindle the interest of students to undertake the management of water and containment of weeds in a scientific way. Questions and answers given at the end of the each chapter can help students to face various competitive exam's in agriculture discipline. I hope that this book will serve as a good text book cum reference material for Undergraduate and Diploma students of agriculture.

I congratulate the authors for having brought out this book covering all the aspects of irrigation and weed management including soil-plant-water relationship, soil moisture constant, soil moisture measurement, irrigation scheduling, different irrigation practices for various crops, micro irrigation technologies, irrigation water quality, irrigation structures, classification of weeds and herblodes, crop weed interactions, principles and methods of weed management and wood management practices for major field crops besides photographic illustrations of woods to enhance the understanding.

(N.KUMAR)

Place: 21.06.2021
Date: Coimbatone

CONTENTS

1

ROLE OF WATER IN PLANTS - IMPORTANCE OF IRRIGATION - WATER RESOURCES OF TAMIL NADU

1. IRRIGATION

Irrigation can be stated as application of water to the soil for crop growth and development. The application of water to plants is made naturally through rainfall and artificially through irrigation.

Irrigation is defined as the artificial application of water to the soil for the purpose of crop growth or crop production in supplement to rainfall and ground water contribution.

1.1 Management

Regulating, the activities based on the various resources for its efficient use and better output. i.e., allocation of all the resources for maximum benefit and to achieve the objectives, without eroding the environment is called management. Otherwise it can be stated as planning, executing, monitoring, evaluating and re-organizing the whole activities to achieve the target.

1.1.1 Irrigation management

Management of water based on the soil and crop environment to obtain better yield by efficient use of water without any damage to the environment.

Management of water, soil, plants, irrigation structure, irrigation reservoirs, environment, social setup and it's inter liked relationship are studied in the irrigation management.

The important aspects in irrigation management are,

- The soil physical and chemical properties.
- Biology of crop plants.
- Quantity of water available.
- Time of application of water.
- Method of application of water.
- Climatological or meteorological influence on irrigation.
- Environment and its changes due to irrigation.

Management of all the above said factors constitute Irrigation Agronomy; Management of irrigation structures, conveyances, reservoirs constitute Irrigation Engineering; and social setup, activities, standard of living, irrigation policies, irrigation association and farmer's participation, cost of irrigation etc., constitute Socio-economic study.

Except Economics and Engineering all the other components are grouped under Agronomy. Sociology has a major role in irrigation management in a large system. Hence Engineering, Economics, Social science and Agronomy are the major faculties come under Irrigation Management.

1.1.2 Importance of Irrigation management

Water is essential not only to meet agricultural needs but also for industrial purposes, power generation, livestock maintenance, rural and domestic needs etc. But the resource is limited and cannot be created as we require. Hence irrigation management it very important:

- To the development of nation through proper management of water resources for the purpose of crop production and other activities such as industrialization, power generation etc., which in turn provides employment opportunities and good living condition of the people.
- To store and to regulate the water resources for further use or non-season use.
- To allocate the water with proper proportion based on area and crop under cultivation. (Balanced equity in distribution).
- To convey the water without much loss through percolation and seepage (Efficiency in use).
- To apply sufficient quantity to field crops. (Optimization of use).
- To utilize the water considering cost-benefit (Economically viable management).
- To distribute the available water without any social problem (Judicial distribution).
- To meet the future requirement for other purposes like domestic use of individual and to protect against famine (Resource conservation).
- To protect the environment from overuse or misuse of water (Environment safe use).

2. IMPORTANCE OF WATER FOR CROP CULTIVATION

Plants or any living organisms cannot live without water, since water is the most important constituent, accounts about 80 to 90% in most of plant cell.

2.1 Role of water in plants

Role of water in plants can be grouped in to two,

2.1.1 Physiological importance

- The plant system itself contains about 90% of water. Amount of water varies in different parts of plant as follows.

Plant portions	Water content
Apical portion of root and shoot	>90%
Stem, leaves and fruits	70-90%
Woods	50-60%
Matured parts	15-20%
Freshly harvested grains	15-20%

Water acts as base material for all metabolic activities:

- All metabolic or biochemical reactions in plant system are driven by water.
- It plays an important role in respiration, transpiration and photosynthesis.
- It activates germination and plays in important role in plant metabolism for vegetative and reproductive growth.
- It serves as a solvent in soil for plant nutrients.
- It also acts as a carrier of plant nutrients from soil to plant system.
- It maintains plant temperature through transpiration.
- It helps to keep the plant erect by maintaining plant's turgidity.
- It helps to transport metabolites from source to sink.

2.1.2 Ecological importance

- Water helps to maintain soil temperature.
- It helps to maintain salt balance.
- It reduces salinity and alkalinity.
- It influences weed growth.
- It influences atmospheric weather.
- It helps the beneficial microbes.
- It influences the pest and diseases.
- It supports human and animal life.
- It helps for land preparation like ploughing, puddling etc.
- It helps to increase the efficiency of cultural operations like weeding, fertilizer application etc. by providing optimum condition.

3. WATER - LIQUID GOLD

The multi-various uses of good quality water for the purpose of irrigation, industrial purposes, power generation, livestock use, domestic use for urban and rural development, are increasing the demand for water. Due to increasing cost of irrigation projects and limited supply of good quality water, it becomes high valuable commodity and hence, it is stated as liquid gold. Further, historical evidences indicate that all civilization established on water base due to proper management and disappear due to improper management of the same water base. All the superior varieties, organic manure, inorganic fertilizer, efficient labour saving implements, better pest and disease management techniques can be implemented only when sufficient water is applied to the crop.

The diversified value of water can be quoted as follows:

- Water as a source of sustenance.
- Water as an instrument of agriculture.
- Water as a community goods.
- Water as mean of transportation.

- Water as an industrial commodity.
- Water as a clean and pure resource.
- Water as a beauty.
- Water as a destructive force to be controlled.
- Water as a fuel for urban development.
- Water as place for recreation and wild life habitat.

4. HISTORY AND DEVELOPMENT OF IRRIGATION IN INDIA

- Irrigation has been practiced since time immemorial, nobody knows when it was started but evidences say that it is the foundation for all civilization since great civilization were started in the river basins of Sind and Nile.
- This civilization came to an end when the irrigation system failed to maintain crop production.
- There are some evidences that during the Vedic period (400 B.C) people used to irrigate their crops with dug well water. Irrigation was gradually developed and extended during the Hindus, Muslims and British periods.
- The Grand Anaicat (KALLANAI) constructed across the river Cauvery is an outstanding example for the irrigation work by a Chola king the great Karikala during second century. The Veeranarayanan Tank and Gangai Konda Cholapuram tank was constructed during 10th century in TN. Anantaraja Sagar in AP was constructed during 13th century.
- Early Mauryan king Samudragupta and Ashoka took great interest in the construction of wells and tanks.
- Later, Moghul kings of North India and Hindu kings of South India focused their attention, in the establishment of canals, dams, tanks etc.
- British Government initiated their work during 19th century in remodeling and renovation of the existing irrigation system. The Upper Ganga canal, Krishna and Godaveri delta system, Mettur and Periyar dams are the great irrigation structures built by the British rulers.
- After independence, Irrigation activities have been accelerated and number of multipurpose river valley projects like Bhakrea- Nangal in Punjab, Tungabhadra in Andraprdesh, Damodar Valley in Madya Pradesh were established.

5. CLASSIFICATION OF IRRIGATION PROJECTS

The irrigation projects can be classified as 1. major 2. medium and 3. minor based on financial limits or expenditure incurred.

1. **Major – more than 50 million Rupees:** It covers cultural command area of more than 10,000 hectares
2. **Medium – 2.5 million to 50 million Rupees:** It covers cultural command area of 2000 – 10,000 hectares
3. **Minor – less than 2.5 million Rupees:** It covers cultural command area of 2,000 hectares.

6. WATER RESOURCES AND IRRIGATION POTENTIAL IN INDIA AND TAMIL NADU

6.1 World scenario

Water is the earth's most abundant component. More than 70% of earth's surface of the world is covered with water. Total water resource in the world is about 1384 million km^3, of which only 26% (360 million km^3) is fresh water and the rest is in ocean (Salt water). Salt content of more than 3% in the sea or ocean is not suitable for agriculture or human consumption. About 77.2% (278 million km^3) of fresh water is in solid form (snow / ice / frost) and rarely used for agriculture. Only 5.2% (18.72 million km^3) is available as surface water and 12.35% (44.46 million km^3) as ground water. Of the total ground water, only about 10% (4.44 million km^3) is easily extractable. Hence, totally, 23.16 million km^3 (18.72 + 4.44) is available for usage in the globe.

In the world, China is having the largest area under irrigation and India stands the second.

6.2 Land resources of India (2014-15)

(Source: Ministry of Agriculture & Farmers Welfare, GOI, ON1700)

Total geographical area	=	328.00 M.ha
Net area sown	=	140.13 M.ha
Total cropped area	=	198.36 M.ha
Net area irrigated	=	68.38 M.ha
Gross area irrigated	=	96.46 M.ha

6.3. India's water scenario

Total geographical area	=	328 M.ha.
Average annual rainfall	=	1190 mm
In million hectare metre	=	1190 × 328 = 392 M.ha.m
Contribution from snowfall	=	8 M.ha.m
Total	=	400 M.ha.m

Rainfall is the ultimate source of all kind of water. The average rainfall of India is 1194 mm. When considered over the geographical area of the 328 M.ha, this rainfall amounts to 392 M.ha.m (or 400 M.ha.m).

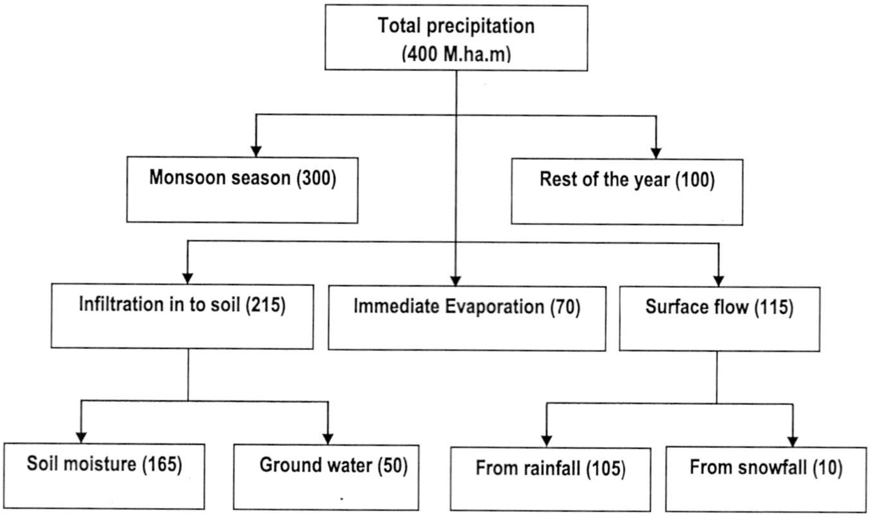

The rainfall below 2.5 mm is not considered for water budgeting, since it will immediately evaporate from surface soil without any contribution to surface water or ground water.

When rainfall occurs, a portion of it immediately evaporates from the ground or transpires from vegetation, a portion infiltrates into the soil and the rest flows over surface as run off.

There are on an average 130 rainy days in a year in the country out of which the rain during 75 days considered as effective rain. The remaining 55 days are very light and shallow which evaporates immediately without any contribution to surface or ground water recharge. Considering all these factors it is estimated that out of 400 M.ha.m of annual rainfall 70 M.ha.m is lost to atmosphere through evaporation and transpiration, about 115 M.ha.m flows as surface run-off and remaining 215 M.ha.m soaks or infiltrates into the soil profile.

6.3.1 Surface water

Includes precipitation (including rainfall and dew), water available from river, tank, pond, lake, etc. Besides, snowfall could able to contribute some quantity of water in heavy snowfall area like Jammu, Kashmir and Himalaya region.

6.3.2 Subsurface water

Includes subsurface water contribution, underground water, well water etc.

6.3.3 Surface run-off

Surface run off consists of direct run off from rainfall, melting of snowfall and flow in streams generated from ground water. Total surface run-off has been estimated by Irrigation Commission of India in 1972 as follows.

(a) Total surface run off	=	180 M.ha.m
(b) Rain fall contribution	=	115 M.ha.m
(c) Contribution from outside the country through streams and rivers	=	20 M.ha.m
(d) Contribution from regeneration from ground water in streams and rivers	=	45 M.ha.m
Total	=	180 M.ha.m

6.3.4 Disposal of surface run off

The surface runoff is disposed in three ways

1. Stored in reservoirs
2. Disappears by means of percolation, seepage and evaporation
3. Goes to sea as waste

The water stored in reservoirs is lost through evaporation and some amount through seepage. The rest in utilized for various purposes mainly for irrigation and drinking.

Total surface run off	=	180 M.ha.m
Stored in reservoir and tanks	=	15 M.ha.m
Flow in the river	=	165 M.ha.m
Utilization from the river by diversion tank and direct pumping	=	15 M.ha.m
Water goes to sea as waste	=	150 M.ha.m
On full development work expected		
Utilization	=	45 M.ha.m
Water flows to sea	=	105 M.ha.m

On full harnessing and mobilization of these water resources, say, by 2025, it is envisaged that 70 m ha m. of surface water and 35 m ha m of ground water can be mobilized for utilization. The projected use of this 105 m ha m of water is 77 m ha m for irrigation and 28 m ha m for domestic and industrial water supply and all other purpose.

6.4 Rain fall

Seasons of rainfall can be classified s follows.

1. Winter (Cold dry period)	—	January - February
2. Summer (Hot weather period	—	March - May
3. *Kharif* (South-West monsoon)	—	June - September
4. *Rabi* (North-East monsoon)	—	October - December

6.4.1 South west monsoon

It comprises the month June, July, August and September which contributes about 70-75% of rainfall to India except for extreme North of Jammu and Kashmir and extremes South of Tamil Nadu. Hence the success of agriculture in India depends on timely onset, adequate amount and even distribution of this South West Monsoon (SWM). This season is also called as *Kharif* season.

6.4.2 North east monsoon

The retreating monsoon comprises the months of October, November and December. North East Monsoon (NEM) contributed rainfall to South Eastern part of peninsular India Tamil Nadu receives its 48% of rainfall. This season is also called as *Rabi* season. About 11% of total rainfall is received during the season.

6.4.3 Winter

It comprises of the month of January and February. It contributes very little rainfall (4-5%) of total in India.

6.4.4 Summer

Comprises of the months of March, April and May and contributes little summer showers. About 10-11% of total rainfall is received during the season.

6.5 Sources of Irrigation in India (2012-13)

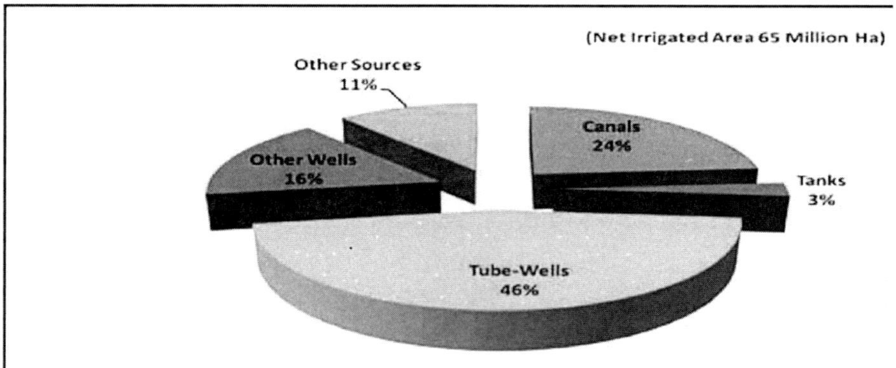

Source: Land Use Statistics (2012-13), Ministry of Agriculture & Farmers Welfare

6.6 Land utilisation pattern in Tamil Nadu

(Source: Season and crop report of Tamil Nadu, 2013-14, Dept. of Economics and Statistics, Chennai - 6)

Total geographical area	=	13.03 M.ha
Gross area under cultivation	=	5.90 M.ha
Net area sown	=	4.71 M.ha
Gross area irrigated	=	3.31 M.ha
Net area irrigated	=	2.68 M.ha

6.7 Rainfall distribution pattern of Tamil Nadu

Season	Normal rainfall (mm)	%
Winter	31.3	3.4
Summer	128.0	13.9
South West Monsoon	321.2	34.9
North East Monsoon	440.4	47.8
Total	**920.9**	**100.0**

6.8 Sources of irrigation in Tamil Nadu (2013-14)

Sources	Area (L ha)	%
Canals (including system tanks)	6.53	24.4
Tanks	3.78	14.1
Wells		
a. Tube wells	4.63	17.3
b. Open wells	11.79	44.0
Others	0.06	0.2
Total	26.79	100.0

(Source: Season and crop report of Tamil Nadu, 2013-14, Dept. of Economics and Statistics, Chennai - 6)

6.9 Area irrigated under different crops in Tamil Nadu (2013-14)

Sl. No.	Crop	Gross area sown (L ha)	Gross area irrigated (L ha)	Gross area irrigated (%)	% of gross area irrigated to gross area sown
1	Paddy	17.26	16.12	48.7	93.4
2	Sorghum	3.47	0.47	1.4	13.5
3	Cumbu	0.54	0.08	0.3	15.4
4	Ragi	1.19	0.33	1.0	27.5
5	Maize	3.80	1.70	5.1	44.7
6	Other cereals	0.33	0.005	0.0	1.7
7	Black gram	3.65	0.63	1.9	17.3

8	Green gram	1.95	0.14	0.4	7.2
9	Horse gram	0.89	0.005	0.0	0.6
10	Other pulses	1.67	0.10	0.3	6.2
	Total food grains	34.75	19.58	59.2	56.4
11	Chillies	0.41	0.15	0.5	36.5
12	Sugarcane	3.13	3.13	9.5	100
13	Other food crops	6.54	3.40	10.3	52.0
	Total food crops	44.83	26.26	79.3	58.6
14	Cotton	1.51	0.37	1.1	24.6
15	Groundnut	3.37	1.38	4.2	41.0
16	Gingelly	0.56	0.21	0.6	37.4
17	Other non - food crops	8.70	4.88	14.7	56.1
Total nonfood crops		14.14	6.84	20.7	48.4
Total food and nonfood crops		58.97	33.10	100.0	56.1

(Source: Season and crop report of Tamil Nadu, 2013-14 of Economics and Statistics, Chennai - 6)

6.10 Irrigation systems of India and Tamil Nadu

State	Project Name
A.P.	Godavari delta system, Krishna delta system, Nagarjuna Sagar (Krishna)
Bihar	Gandala
Punjab	Western Jamuna, Bhakra-Nangal, Sutlej, Beas
Gujarat	Kakrapare – Tati, Narmada
M.P.	Gandhi sagar (Chambal, Ranap setab, sagar
Maharastra	Bhima Jayakwadi (Godavari)
Kerala	Kalada, Mullai Periyar
Karnataka	Ghataprabha, Malaprapha and Turga
Orissa	Hirkand and Mahanathi
U.P	Upper Ganga canal, Ramaganga
W.B	Damodar Valley
Rajasthan	Rajasthan Canal (Sutlej)
Tamil Nadu	Mettur – Lower Bhavani Project, Parambikulam Alliyar Project, Periyar Vaigai, Tamiarabarani

QUESTIONS

I. Fill in the blanks

1. Plant cell contains _______ to _______ % of water
2. The Grand Anaicat (KALLANAI) constructed across the river Cauvery by _______ king during ________century
3. Fresh water of the world is ________million km^3
4. Average annual rainfall India is _______ mm
5. North East monsoon is otherwise called as _______ monsoon
6. Largest irrigation canal in India is ____________

II. Choose the best answers

1. ________ acts as base material for all metabolic activities
   (a) Air (b) Nutrients (c) Water (d) Light
2. Total water resource in the world is about _________ million km^3
   (a) 1384 (b) 1500 (c) 900 (d) 1200
3. Net irrigated area of India is ______ m ha during 2014-15
   (a) 68.38 (b) 98.10 (c) 45.09 (d) 102.25
4. South west monsoon contributes about ________% of rainfall to India except for extreme North of Jammu and Kashmir
   (a) 30 (d) 45 (c) 60 (d) 75
5. Net irrigated area of Tamil Nadu is _____ m ha during 2013-14
   (a) 6.28 (b) 2.68 (c) 4.5 (d) 10.12

Answer the following questions

1. Define: Irrigation
2. What is irrigation management?
3. Explain the role of water in plant system
4. Importance of irrigation projects in agriculture
5. List out major irrigation projects of Tamil Nadu

2

SOIL MOISTURE CONSTANTS-PLANT WATER STRESS-PLANT RESPONSE AND ADAPTATIONS-METHODS TO OVERCOME PLANT WATER STRESS

Soil is a complex system made of solid, liquid and gaseous materials.Soil can be defined as a complex, dynamic, three phase system composed of solid, liquid and gas with some proportions. Normally the proportion is 50:25:25 but this may vary among soils and in some cases one of the components may be absent.Water logged soil air is not present and desert sandy soils water component may not be available. Soil serves as a reservoir for moisture and knowledge about soil water relationship is the principal factor governing the frequency and amount of irrigation water to be applied to the crop.

Solid

The solid phase is made of minerals, organic matter and various chemical compounds.

Liquid

The liquid portion of soil consists of water, dissolved minerals and soluble organic matter. Soil water is stored in the space between soil particles known as pore space and this pore space is the most important physical structure and plays vital role in irrigation management.

Gas

The space in between soil particles is not only filled with water, but some spaces are occupied with air. The pore spaces enclosed by soil matrix is shared by soil air and soil water. Soil air contains lesser oxygen and more carbon di oxide than atmosphere air due to respiration by soil microbes and plant roots in which oxygen is consumed and carbon di oxide is released.

Soil water or soil moisture

The water present in the soil pore spaces is the most important soil ingredient and plays a vital role in crop production. The water received after rain or irrigation is stored in the soil profile in pore spaces and the stored soil water is known as soil moisture.

Water is retained as thin film around the soil particles and in the capillary pores by the forces of adhesion, cohesion and surface tension.

Adhesion

Adhesion is the force of attraction between molecules of different substances, the force of attraction between soil mass to liquid surface. A thin film of water is held in soil particles due to this adhesive force.

Cohesion

Cohesion is the force of attraction between molecules of same substances i.e. between water molecules.Hence a thick film of water is formed due to this cohesive force.

Surface tension

It is the total force acting in a solid, liquid, air system. As a result of this, the soil - air inter space become curved.

Soil moisture tension

Soil moisture tension is the measure of the tenacity with which water is held in the soil and shows the force per unit area required to remove water from the soil.

This pressure or tenacity is measured in terms of potential energy of water and expressed in equiatmosphere or bars.

1 atmosphere = 1036 cm water column (or)
76.39 cm of mercury

1 bar = 1023 cm water column

To convert the soil moisture tension to equivalent atmosphere the above conversion ratio can be used.

Solute potential

Portion of water potential that results from the solute present in the soil systemsis known as solute potential. Sometimes it may called as osmotic potential.

1. SOIL MOISTURE CONSTANTS

The water contents expressed under certain standard conditions are commonly referred to as soil moisture constants. They are used as reference points for practical irrigation water management. The usage of these constants together with the energy status of soil water gives useful knowledge.

Soil moisture constants refer to the status of the soil mass or changes occurring in the soil mass after irrigation or rainfall. In the real sense, soil moisture cannot be said to be constant since it is very dynamic and always tends to change due to potential gradient or pressure gradient. This phenomenon helps to find out the soil moisture status, availability of soil moisture, time and quantity of irrigation water to be applied etc.

The soil moisture constants are

1. Saturation capacity or maximum water holding capacity (MWHC)
2. Field capacity (FC)
3. Permanent wilting point (PWP)
4. Available soil moisture (ASM)
5. Moisture equivalent
6. Hygroscopic coefficient

1.1 Saturation capacity or maximum water holding capacity

Saturation capacity refers to the condition of soil at which all the macro and micro pores are filled with water and the soil is at maximum water retention capacity (Fig.1). The matric suction at this condition is essentially zero as the water is in equilibrium with free water. Excess water above saturation capacity of soil is lost from root zone as gravitational water.

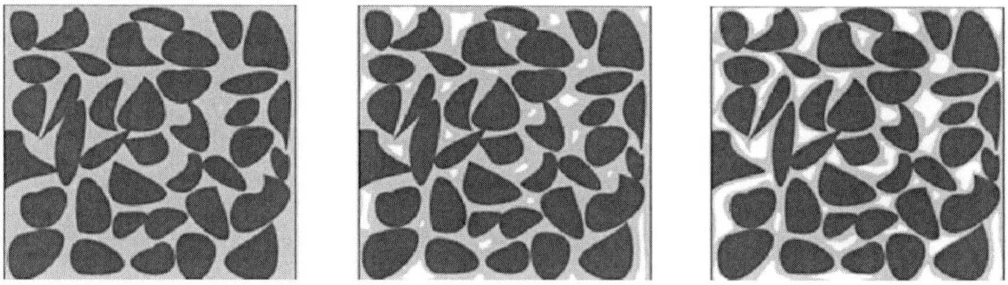

Fig. 1. Soil condition at Saturation (a), Field Capacity (b) and PWP (c)

1.2 Field Capacity (FC)

According to Veihmeyer and Hendrickson (1950) the field capacity is the amount of water held in soil after excess water has been drained away and the rate of downward movement has materially decreased, which usually takes place within 1-3 days after a rain or irrigation in pervious soils having uniform texture and structure (Fig.1). At field capacity, the soil moisture tension depending on the soil texture ranges from 0.10 to 0.33 bars (or −10 to −33 kPa). Field capacity is considered as the upper limit of available soil moisture. The field capacity is greatly influenced by the size of the soil particles (soil texture), finer the soil particles higher the water retention due to very large surface area and vice versa. Thus, at field capacity, a m^3 of a typical sandy soil will hold about 135 litres of water, a loamy soil about 270 litres and a clay soil about 400 litres.

1.3 Permanent Wilting Point (PWP)

It is the condition of the soil wherein water is held so tightly by the soil particles that the plant roots can no longer obtain enough water at a sufficiently rapid rate to satisfy the transpiration needs to prevent the leaves from wilting. When this condition is reached the soil is said to be in a state of permanent wilting point, at which nearly all the plants growing on such soil show wilting symptoms and do not revive in a dark humid chamber unless water is supplied from an external source (Fig.1). The soil moisture tension at permanent wilting point is about 15 bars. Permanent wilting point is considered as lower limit of available soil moisture. Under field conditions, PWP is determined by growing indicator plants such as sunflower in small containers. In the laboratory, pressure membrane apparatus can be used to determine the moisture content at 15 bars.

1.4 Available Soil Moisture (ASM)

It has been a convention and even now it is a customary to consider the amount of soil moisture held between the two cardinal points viz., field capacity (0.33 bars) and permanent wilting point (15 bars) as available soil moisture (Fig.1). Though considerable soil moisture is present below the permanent wilting point, it is held so tightly by the soil particles that the plant roots are unable to extract it rapidly enough to prevent wilting. Thus, practically it is not useful for the plants and forms the lower limit of available soil moisture. Similarly, the water above the field capacity is not available to the plants owing to quick drainage.

Table 1. Moisture holding properties of different soils varying in texture

Soil texture	Saturation capacity (%)	Field capacity (%)	Permanent wilting point (%)	ASM		Infiltration rate (mm/hr)
				%	mm/m	
Clayey	60	40	20	20	200	3
Clay loamy	50	30	15	15	150	3-7
Silt loamy	45	22	12	10	100	7-12
Sandy loam	42	14	6	8	80	12-20
Loamy sand	40	10	4	6	60	20-30
Sandy	38	6	2	4	40	30

1.4.1 Factors affecting available water range

The role of soil texture, organic matter content and the amount of cementing material in increasing water retention of soil is well established. It is generally Accepted that at any given moisture retention value, coarse textured soils hold less water than medium textured soils and these in true less water than fine textured soils. Soil structure influences the rate at which water and air enter and move through the soil profile as these are closely related to pore. Since organic matter has greater water holding capacity than mineral soils, it is believed addition of organic matter soil result in increasing water holding capacity of the latter. Increase in the water holding properties and available water range have been attributed to the increase in fine particles than to the organic matter, which may not change materially due to its fast rate of decomposition under tropical condition. Increase concentration of $CaCO_3$ and Iron oxide (FeO) decrease the moisture holding capacity of soil. The osmotic potential of soil solution at field capacity is about -1/10 bar in normal soils but it may be -4 to-6 bars or even higher in very saline soils. Plants under this severe condition may wilt at moisture contents appreciably above the wilting point of the same soil with lower salt content.

1.4.2. Concepts of soil – water availability to plants

Three concepts of soil –water availability to plants are, generally cited in literature;

- Water is equally available to plants from FC to PWP
- Availability decreases with decreasing soil wetness from FC to PWP
- Readily available up to a critical point between FC and PWP followed by gradual decrease.

Optimum plant growth and development normally take place at field capacity moisture content. It is not known whether the water is equally available for plant growth over the entire available soil moisture range. There are three theories of soil water availability to plants, as follows (Fig.2)

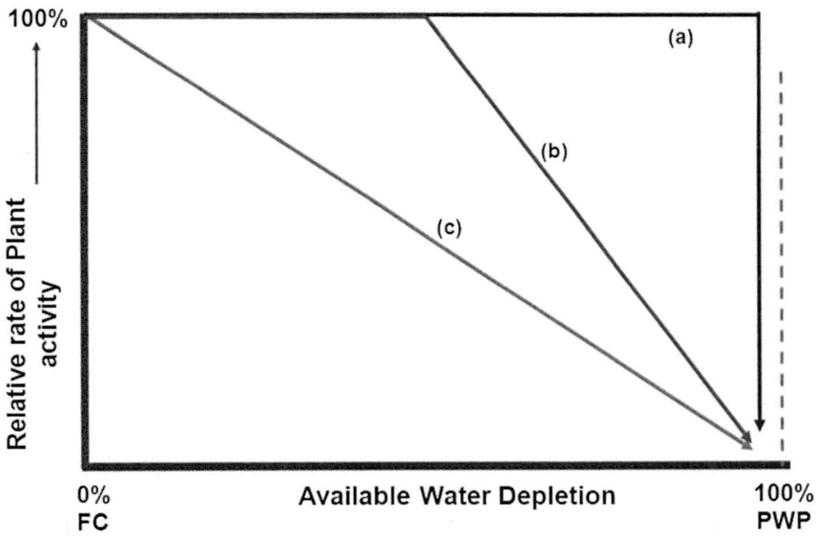

Veihmeyer and Hendrickson (1950) proposed that soil water is equally available to plants equally throughout a definable range of soil wetness, from an upper limit FC to a lower limit PWP, both of which are characteristic and constant for a given soil (Curve '*a*'). According to this theory, plant functions remain unaffected by any decrease in soil wetness until PWP is reached, at which plant activity is curtailed, often abruptly.

Other investigators notably Richards and Wadleigh (1952) indicated that soil water availability to plants actually decreases with decreasing soil wetness and that a plant may suffer water stress before wilting is reached (Curve '*c*').

Still others attempted to divide available range of soil wetness into readily available and decreasingly available zones and searched for a critical level somewhere between FC and PWP as an additional criterion of soil water availability (Curve '*b*').

1.5 Hygroscopic coefficient

It is defined as the amount of water that the soil contains when it is in equilibrium with air at standard atmosphere i.e., 98% relative humidity and at room temperature. In other words, it is the amount of moisture absorbed by a dry soil when placed in contact with an atmosphere saturated with water vapour (100% relative humidity) at any given temperature, expressed in terms of percentage on an oven dry basis. The matric suction of soil water at this moisture content is nearly about 31 bars.

1.6 Moisture equivalent (ME)

Moisture equivalent is defined as the amount of water retained by a sample of initially saturated soil material after being subjected to a centrifugal force of 1000 times that of gravity for a definite period of time, usually half an hour.

In medium textured soils, values of field capacity and moisture equivalent are nearly equal. In sandy soils, the FC exceeds the ME and clay soils FC is lower than ME. To estimate moisture equivalent, a small mass of soil sample is saturated with water and is subjected to centrifugal force of 1000 times that of gravity for half an hour and the soil moisture percentage is worked out by gravimetric method. This moisture percentage is equal to moisture equivalent.

1.7 Importance of soil moisture constants in irrigation management

The concept of soil moisture constants greatly facilitates, in taking decisions in irrigation water management. According to soil water regime concept, the water concept at FC was considered as upper limit (100%) of available water for crop growth and that at PWP is lower limit (0%). However, entire ASM is not equally to crop growth. There is a decreasing availability of water for crop growth, before PWP has been reached. Hence, the deficit in ASM (25% or 50 % DASM), at which irrigation to be given is pre determined first. It is assumed that level, growth of the crop was likely to be reduced. Hence, the crop is irrigated, when the soil moisture in the root zone is depleted to a particular level. Therefore, net irrigation (NIR) is the difference between the field capacity and soil moisture content in the root zone before starting irrigation. In other words, it is the amount of irrigation water required to bring the soil moisture in the effective root zone depth to field capacity. Gross irrigation requirement (GIR) is the NIR plus losses in water application and other losses.

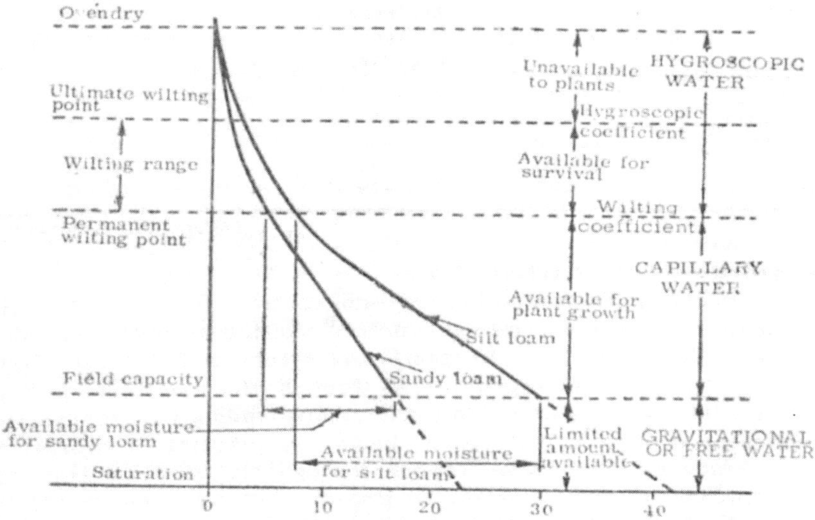

Fig. 3. Kinds of soil water and available moisture content in sandy and silt loam soil

2. SOIL WATER MOVEMENT

2.1 Infiltration

Infiltration is the entry of fluid from one medium to another. In irrigation practice, it is the process of water entry into the soil, generally by downward flow through all or part of the soil surface.

Infiltration rate or infiltrability is defined as the volume of water flowing into the profile per unit of soil surface area per unit time.

It is mathematically expressed as:

$$i = \frac{Q}{A \times T}$$

Where,

i = Infiltration rate (mm or cm/min or h)

Q = Volume of water (m^3) infiltrating

A = Area of the soil surface (m^2) exposed to infiltration

T = Time (min or h).

The infiltration rate is not constant over time. Generally, infiltration rate is high in the initial stages of infiltration process, particularly where the soil is quite dry, but tends to decrease monotonically and eventually to approach asymptotically a constant rate, which is often termed as basic intake rate or steady state infiltration rate. Whereas, the cumulative infiltration, being the time integral of the infiltration rate, has curvilinear time dependence, with a gradually decreasing slope as shown in Fig. The infiltration rate of a soil may be easily measured using a simple device known as a double ring infiltrometer placing *in-situ*.

The variation of infiltration rate in different soil textures is shown in the table. The factors influencing infiltration rate are time from the onset of rain or irrigation, initial water content, hydraulic conductivity, presence of impeding layers in the profile and vegetative cover.

Table 2. Steady state infiltration rates of various soil types

Soil	Infiltration rate (mm/hr)	Soil	Infiltration rate (mm/hr)
Sandy	30	Silt loam	7-2
Loamy sand	20-30	Clay loam	3-7
Sandy loam	12-20	Clayey	< 3

2.2 Seepage

The lateral movement of water through soil pores or small cracks in the soil profile under unsaturated condition is known as seepage.

2.3 Permeability

It indicates the relative ease with which air and water penetrate or pass through the soil pores. Permeability of soils is generally classified as rapid, moderate and slow. Thus, the permeability is rapid in coarse textured soils and slow in fine textured soils.

2.4 Deep percolation

Infiltration is a transitional phenomenon that takes place at the soil surface. Once the water has infiltrated the soil, the water moves downward into the profile. This post infiltration water movement downward within the soil profile under the influence of both gravity and hydrostatic pressure is termed as deep percolation. Sandy soils facilitate greater percolation when compared to clayey soils due to dominance of macro pores. Likewise, the loss of water by percolation in cropped fields is generally less than that in bare soils.

2.5 Types of water movement

Movement of water in the soil is complex because of various states and directions in which water moves and because of the forces that cause it to move. As water is dynamic soil component, generally three types of water movement within the soil are recognized – saturated flow, unsaturated flow and water vapour movement. The former two flows involve liquid water in contrast to water vapour flow. In all cases, water flows in response to energy gradients, with water moving from a zone of higher to one of lower water potential.

2.5.1 Saturated water flow

The condition of the soil when all the macro and micro pores are filled with water the soil is said to be at saturation, and any water flow under this soil condition is referred to as saturated flow. The flow of water under saturated conditions is determined by two factors viz., the hydraulic force driving the water through the soil (hydraulic gradient) and the ease with which the soil pores permit water movement (hydraulic conductivity). The quantity of water per unit of time (Q/t) that flows through a column of saturated soil can be expressed by Darcy's law, as follows:

$$\frac{Q}{t} = AK_{sat}\frac{\Delta\Psi}{L}$$

Where,

A = Cross-sectional area of the column through which the water flows

K_{sat} = Saturated hydraulic conductivity

$\Delta\Psi$ = Change in water potential between the ends of the column (for example, $\Psi_1 - \Psi_2$)

L = Length of column

For a given column, the rate of flow is determined by the ease with which the soil transmits water (K_{sat}) and the amount of force driving the water, namely the water potential gradient ($\Delta\Psi/L$). For saturated flow, this force may also be called the hydraulic gradient. By analogy, think of pumping water through a garden hose, with K_{sat} representing the size of the hose (water flows more readily through a larger hose) and $\Delta\Psi/L$ representing the size of the pump that drives the water through the hose. The units in which K sat is measured are length/time, typically cm/s or cm/h.Generally the rate of flow is in the following sequence: Sand > loam >clay.

2.5.2 Unsaturated water flow

The soil is said to be under unsaturated condition when the soil macro pores are mostly filled with air and the micro pores (capillary pores) with water and some air, and any water movement or flow taking place under this soil condition is referred to as unsaturated flow. Under field conditions, most of the soil water movements occurs only when the soil pores are not completely saturated with water. However, water movement under these conditions is very sluggish compared to that when the soil is at saturation. This is because at or near zero tension, the tension at which saturated flow occurs the hydraulic conductivity is orders of magnitude greater than at tensions of 0.1 bars and above, which characterize unsaturated flow. In unsaturated condition, the conductivity is more in fine soil than coarse textured soil. The rate of unsaturated flow is in the following order: Sand < loam < clay.

2.5.3 Vapour movement

Two types of vapour movement occurs in soil; internal and external. Due to differences in vapour pressure from one area to another in the soil internal movement of vapour take place within the soil. External vapour movements take place on the surface of the soil due to surface evaporation of soil water. Since water in the form of vapour in the soil is small its movement in soils is not of practical importance except in dry soils for drought resistant desert plants.

Particulars	Saturated flow	Unsaturated flow	Vapour movement
Major force	Gravitational	Metric	Vapour pressure
Water form	Liquid	Liquid	Vapour
Major direction of flow	Downward	Lateral	All direction
Pore space	All pores filled with water	Micropores filled with water	All pores are empty
Rate of flow	Fast(1cm-100cm\ day)	Slow(0.01cm to 0.00001cm/day)	-
Volume of water movement	Large quantities (3,75,000kg/ha in 15cm depth)	Small (1,00,000kg/ ha in 15cm depth of soil)	Negligible 15kg/ ha in 15cm depth of soil)

3. PLANT WATER RELATIONS

Almost every plant process is affected by water supply either directly or indirectly. Water forms a continuous liquid phase from the root hairs to the leaf epidermis.Plant roots take up water through active and passive absorption process and transfer it through the root tissues to the conducting vessels (xylem) of the plants. After moving through these tissues, the water reaches the evaporation sites (the walls of sub stomatal cavity).All these flows are in the liquid phase along a potential gradient.The final transfer of water from the sub-stomatal cavities through the stomata is a process of molecular diffusion of water vapour.

The factors influencing the water relation of plants and in turn the growth and yield are:

1. Soil factors = Soil moisture content, texture, structure, density, salinity, fertility, aeration, temperature and drainage
2. Plant factors = Type of crop, density and depth of rooting, rate of root growth, drought tolerance etc.
3. Weather factors = Sunshine, temperature, humidity, wind and rainfall
4. Miscellaneous = Soil volume and plant spacing, soil fertility, crop and soil Management

The three main areas of water plant relationship are:

1. Water absorption
2. Water conductance and translocation
3. Water loss or transpiration

These three processes have to be clearly understood to analyse the effect of these processes on plant growth and crop yield and in order to identify / formulate steps which are needed to regulate and modify the cropping system to obtain maximum water use efficiency.

4. SOIL-PLANT-WATER RELATIONSHIP

For designing, operating and managing any irrigation system in efficient manner, knowledge of movement, retention and use of water is necessary. Soil-plant-water relationships have to be understood to plan an irrigation system to suit different crops grown on different soils.

Soil-plant-water relations mainly governed by rooting characteristic of plants, effective root zone depth of plant and moisture extraction pattern of plants.

4.1 Rooting characteristic of plants

The purpose of irrigation is to provide adequate soil moisture in the immediate vicinity of the plant roots. All plants do not have the similar rooting pattern i.e., root penetration and proliferation. Some plants have relatively shallow root system (annual crops), while others develop several metres under favourable conditions (tree crops). It is obvious, therefore, that if one is to plan an efficient irrigation schedule for providing adequate soil moisture to plant roots, it is necessary to understand rooting habits of plants.

4.2 Effective root zone depth

The soil depth from which the crop extracts most (nearly 90%) of the water needed to meet its evapotranspiration requirements is known as effective crop root zone depth. It is also referred to as design moisture extraction depth, the soil depth used taken into account to determine the irrigation water requirements for scheduling irrigation of crops.

4.2.1 Rooting depth of crop

Shallow rooted crops (60cm): Cauliflower, Cabbage, Onion, Potato, Lettuce, Rice

Moderately deep rooted (90cm): Carrots, French bean, Garden pea, Chilli, Muskmelon, Tobacco, Wheat, Castor, Banana, Groundnut

Deep rooted (120cm): Cotton, Tomato, Watermelon, Maize, Sorghum, Sugarbeet, Soybeans, Pearl millet

Very deep rooted (180cm): Lucerne, Citrus, Apple, Guava, Grapevine, Coffee, Tea, Sugarcane, Safflower, Mango, Pomegranate

4.3 Moisture extraction pattern

The moisture extraction pattern refers to the amount of soil moisture expressed as percentage extracted from different layers in the soil profile. In most plants, the concentration of absorbing roots is the greatest in the upper part of the root zone (usually in the top 45cm) and near the base of the plant. Hence, extraction of water from the topsoil layers is usually more compared to lower layers. Since water also evaporates from upper few inches of soil, it is lost rapidly from the upper layers. As the amount of moisture in the upper part of root zone in the vicinity of roots is diminished, soil moisture tension increases and a moisture suction gradient is created between the upper layers and the far away lower layer from where moisture has not been extracted previously. This causes water to move towards the upper layers to attain equilibrium between two zones, thus, the plants get moisture from lower parts of the root zone. The usual extraction pattern (Fig.4) shows that about 40% of the soil moisture extracted is contributed from upper quarter of the root zone, 30% from the second quarter, 20% from the third quarter, and 10% from the bottom quarter. Values for individual crops may vary within the range of 10%.

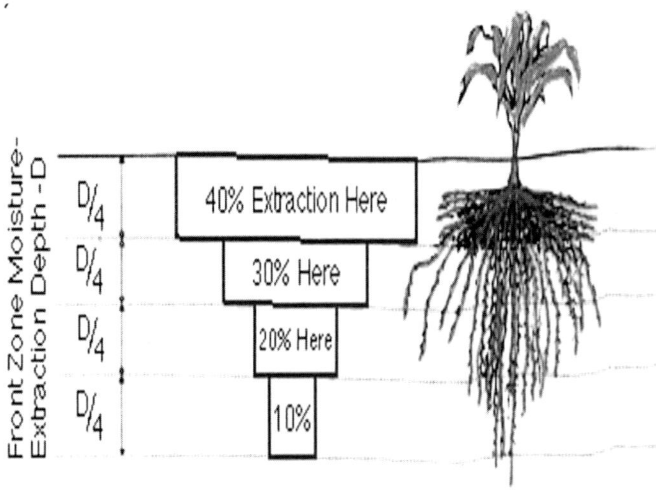

Fig. 4. Moisture extraction pattern

5. SOIL-PLANT-ATMOSPHERE CONTINUUM (SPAC)

The path of water which moves from the soil through the plant to the atmosphere forms a continuous system. The soil-plant-atmosphere continuum (SPAC) refers to the many inter-connections between the physical and organic components of the earth.

The path of the above may be divided into four sequential processes as follows.

1. Supply of liquid water to the root surface
2. The entry of water into the root
3. The passage of water in the conducting elements
4. The movement of water through and out of the leaves

The rate of water movement is always proportional to the potential gradient and inversely proportional to the resistance to flow. The flow rate of water in the soil-plant-atmosphere system is given by the following equation.

$$\text{Flow rate} = \frac{\psi\ \text{soil} - \psi\ \text{root surface}\ \psi}{y\ \text{soil}} = \frac{\text{root surface} - \psi\ \text{xylem}}{y\ \text{root}}$$

$$= \frac{\psi\ \text{xylem} - \psi\ \text{leaf}\ \psi}{y\ \text{xylem} + y\ \text{leaf}} = \frac{\text{lear} - \psi\ \text{air}}{y\ \text{leaf} + y\ \text{air}}$$

In which ψ is the water potential at various sites of system and γ is the corresponding resistance.

The relative magnitudes of resistances depend on soil, plant and atmospheric conditions. Lower the water potential, higher the resistance. Plant offers little or no resistance when soil has sufficient soil moisture and atmospheric conditions are moderate. When soil dries, water deficits develop in plant and stomata closes partially or completely under these conditions plants offer greater resistance for water movement. In the SPAC water moves from higher potential to lower potential. Water potential of the soil near field capacity is 0.01 to 0.03 Mpa and near permanent wilting point is –1.5 Mpa. The normal growing plants show water potential of –0.4 to –1.5 Mpa. Water potential of atmosphere in a humid weather is –10 Mpa and in a dry weather it is –100 Mpa. On a clear day in the tropics and with sufficient soil moisture, the probable water potentials in the soil, plant and atmosphere are –0.05 Mpa, –0.5 Mpa, –100 Mpa respectively. Due to this steep gradient in water potential from soil into atmosphere, water moves from soil into the atmosphere.

6. ABSORBTION OF WATER AND EVAPOTRANSPIRATION

Evapotranspiration (ET) = Evaporation + Transpiration

Evaporation is a diffusive process by which water from natural surfaces, such as free water surface, bare soil, from live or dead vegetation foliage (intercepted water, dewfall, guttation etc.) is lost in the form of vapour to the atmosphere. It is one of the basic components of hydrologic cycle. Likewise, transpiration is a process by which water is lost in the form of vapour through plant surfaces, particularly leaves. In this process, water is essentially absorbed by the plant roots due to water potential gradients and it moves upward through the stem and is ultimately lost into the atmosphere through numerous minute stomata in the plant leaves. It is basically an evaporation process. However, unlike evaporation from a water or soil surface, plant structure and stomatal behaviour operating in conjunction with the physical principles governing evaporation modify transpiration.

6.1 Evaporation

A physical process in which liquid water is converted into its vapour. Evaporation is the most important water loss term in water balance equation.

The essential requirement for evaporation process are

1. Source of heat energy to vapourise the irrigated water
2. Presence of a concentration gradient of water vapour between the evaporating surface and surrounding air or atmosphere.

Evaporation occurs only when vapour concentration of evaporating surface exceeds that of surrounding air.Evaporation from land surface is affected by degree of wetness of the surface, temperature of air and soil, atmospheric humidity, wind velocity and vegetative cover.

6.2 Transpiration

This is a physiological phenomenon, which takes place only in living plants. The loss of water from living parts of the plant is known as Transpiration. The loss of water through stomatal openings of the leaves is termed as stomatal transpiration. The loss of water through cuticle is known as cuticular transpiration and from lenticels is known as lenticular transpiration.

Table 3. Differences between evaporation and transpiration

Evaporation	Transpiration
Controlled by meteorological factors	Controlled by both meteorological and plant factors
Diffusive resistant is absent	Diffusive resistance occurs due to internal leaf geometry and presence of stomata.
Also occurs in night due to advective heat transportation.	Reduced in the night due to closure of stomata.
This is purely a physical phenomenon which takes place from any exposed surface with moisture.	This is physiological phenomenon which takes place only in living plants.
This takes place through any openings or pores	This takes place through guard cells of stomata, cuticle and lenticules etc.

6.3 Evapotranspiration (ET)

Evapotranspiration denotes the quantity of water transpired by plants or retained in the plant tissue plus the moisture evaporated from the surface of the soil. As long as the rate of root uptake of soil moisture balances the water losses from the canopy, evapotranspiration continues to occur at its potential rate. When the rate of root water uptake falls below the transpiration demand, actual transpiration begins to fall below the potential rate. This is either because the soil cannot supply water to roots quickly or the plant can no longer extract water to meet the evaporational demand.

6.3.1 Factors affecting ET

1. Solar radiation which supplies energy for ET
2. Wind which removes the water vapour from cropped area and makes changes in water vapour concentration gradient.
3. Temperature which increases ET rate
4. Relative humidity which changes the ET rate due to changes in water vapour gradient
5. Stage of the crop

   This is very particular in annual crops which have distinct growth stages.

   (a) Emergence and development

   ET or CU increases rapidly from low value and approaches maximum

(b) Maximum vegetative phase

ET or CU reaches maximum if abundant soil moisture is available

(c) Maturity phase

ET or CU begins to decrease

6. Rooting characters of plants
7. Environment
8. Management practices including crop geometry, density, intercropping, mulching and weeding operations.

6.4 Reference evapotranspiration (ET_0)

This represents the maximum rate of evapotranspiration from an extended surface of 8 to 10 centimeters tall green grass cover, actually growing and completely shading the ground under limited supply of water.

6.5 Potential evapotranspiration (PET)

Potential evapotranspiration (PET) for any crop is obtained from reference evapotranspiration and crop factors (*Kc*) when water supply is unlimited.

$$\text{PET} = Kc \times ET_0$$

6.6 Importance of evapotranspiration and potential evapotranspiration for crop plants

Estimation of the soil moisture there by planning irrigation schedule of crops. Understanding relationship between the crop yield and irrigation water. Guiding for the production of a crop with a fully developed canopy. The evapotranspiration can also help to demarcate soil climatic zones including the drought prone areas. These will form the base for developing suitable soil and crop management practices, crop varieties, water conservation techniques, cropping pattern and ways to improve productivity rainfed crops.

7. SOIL MOISTURE ESTIMATION

The measurement of soil moisture is needed to determine when to irrigate and the amount of water needed when irrigating, to evaluate evapotranspiration, and to monitor soil matric potential. The soil moisture is measured in two ways – direct and indirect methods as follows:

Direct methods: Gravimetric and Volumetric method, Spirit burning method, Infrared moisture balance.

Indirect methods: Resistance blocks, Neutron probe, Tensiometer, Pressure plate and membrane.

7.1 Direct methods

The principle involved is the amount of moisture held by a given mass or volume of soil can be expressed as weight or volume basis in percentage. Soil moisture on weight basis is based on the dry weight of the sample.

Soil moisture measurements should be made in the effective root zone depth of the particular crop. For row crops, the measurements should be made in the row or near the plants.In sprinkler irrigation, measuring place should be between the sprinkler heads and 3-4 m away from the laterals.Soil sampling tools like screw auger, core sampler, post-hole auger, tube auger and bucket type auger are required for sampling.

7.1.1 Soil moisture estimation by gravimetric method

Materials required in this method are screw auger, aluminium moisture cans, metal rod, gunny bag, plastic sheet, hot air oven, weighing balance.

Procedure

Drive the screw auger to different depths of soil (eg.0-15, 15-30, 30-45 cm etc.)Take a composite sample of about 50 g in an aluminium can and cover it immediately with lid. Cover the cans with a plastic sheet/gunny bags to avoid loss of moisture. Take the initial weight of the sample on a balance (W_1). Dry the sample in an oven to a constant weight at 105°C (W_2) (takes about 48 hours). Calculate the moisture percentage by the following formula.

$$\text{Moisture percentage} = \frac{W_1 - W_2}{W_2} \times 100$$

Where,

W_1 - initial weight of the sample (g)

W_2 - Oven dry weight of the sample (g)

Merits	Demerits
• Used mostly for experimental purposes	• Laborious
• Accurate measurement	• Time consuming

7.1.2 Soil moisture estimation by volumetric method

$$\text{Soil moisture} = \frac{\text{Weight of moist sample} - \text{Weight of oven dry sample}}{\text{Weight of oven dry sample}} \times 100 \times \text{Bulk Density}$$

The per cent by volume of moisture content obtained by the above relationship is numerically equal to cm of water per metre depth of soil depth.For example, if the bulk density is 1.6 gm/cc and the moisture content is 15 per cent, on dry weight basis, the moisture is 1.6 × 15 = 24 cm per metre depth of soil.

7.1.3 Spirit burning method

Soil moisture from the sample is evaporated by adding alcohol and igniting. Provided, the sample is not too large, the result can be obtained in less than 10 minutes. About 1.0 ml of spirit or alcohol per g of soil sample at field capacity and 0.5 ml at permanent wilting point is adequate for evaporating the soil moisture. This method is not recommended for soils with high organic matter content.

7.1.4 Infrared moisture balance

It consists of a 250 watt infrared lamp, sensitive balance and autotransformer. All housed in an aluminium cabinet. The radiation emitted by infrared lamp quickly vaporizes the soil moisture. The instrument is directly calibrated in per cent moisture. It gives fairly reliable moisture estimates in about 5 minutes.

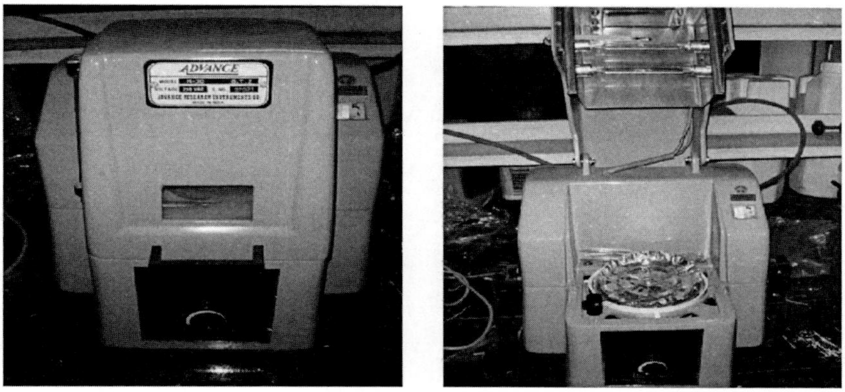

Fig. 5. Infrared moisture balance

7.2 Indirect methods

7.2.1 Resistance blocks

Gypsum blocks, plaster of paris, fibre glass are used for making resistance blocks and are used for measurement of soil moisture *in situ*. It is an indirect measurement of soil water content.

Resistance block works on the principle of conductivity of electricity. When two electrodes *A* and *B* are placed parallel to each other in a medium and when electric current is passed, the resistance to the flow of electricity is proportional to the moisture content in the medium. Thus, when the block is wet, conductivity is high and the resistance is low. The reading at field capacity varies from 400-600 Ohms and 50000 to 75000 Ohms at wilting point. The readings are taken with a portable Wheat stone bridge operated by dry cells.

Installation of gypsum blocks

Materials required are gypsum or nylon blocks and posthole auger.

Procedure

Calibration of the resistance blocks are first step.Sink a bore with a posthole auger to the desired depth. Place the blocks inside and fill back the bore in small depths by tampering the soil with a metal rod. Ensure an intimate contact of the block with the soil. There should not be any root pieces, pebbles etc. near the blocks. Normally 4-5 blocks can be placed in one bore at a vertical interval of 30 cm. Heap the soil to a height of about 3 cm near the surface at the bore spot to prevent any water stagnation. Irrigate the field and record the readings. Check the resistance reading at field capacity. Install the block in a row in between two plants. The resistance blocks read low resistance at field capacity and high resistance at wilting point.

Merits

- Works at low moisture levels up to wilting point.
- Suitable for repeated measurement at a point.

Limitations

- Gypsum blocks are sensitive to salinity and soil temperature.
- They have short life in wet soil.
- Calibration is required for each soil.
- Not suited to high moisture conditions.

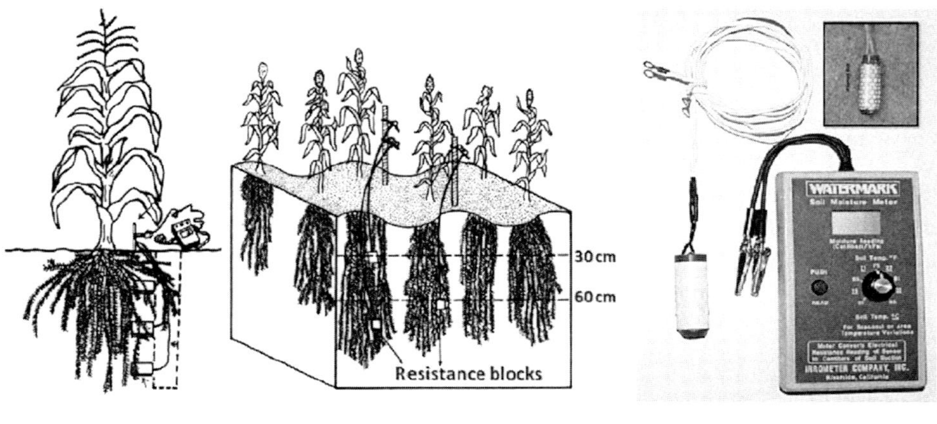

Fig. 6. Measurement of soil moisture by resistance blocks

7.2.2 Soil moisture estimation by neutron moisture meter

The neutron moisture meter is a device to measure the soil moisture status (volumetric water content) *in situ* without disturbing the soil system. Hydrogen nuclei have a property for scattering and slowing neutrons and the phenomenon of the loss of maximum kinetic energy when neutrons collide with hydrogen nuclei of water forms the basis of soil moisture estimation. Neutron source (radium beryllium (5 MC) or americium beryllium (30 MC), shield for storage of the neutron source, detector of slow neutrons, counting device (Scaler) and access tube (GI pipe of 50 mm diameter) are required.

Procedure

The access tube is first inserted into the soil after drilling a hole with the help of an auger. The access tube is kept few inches above the soil.Insert the neutron probe in the access tube at the desired depth. While making a measurement, turn on the scaler a few minutes earlier to warm up. Make several standardization counts with the probe in the shield without altering its position. Then take readings at successive depth intervals.Divide readings by the standard reading to obtain a count ratio and refer to the instrument calibration curve to obtain water content by volume at various depths. Cover the access tube with a lid after taking readings to avoid entry of water in to the tube.

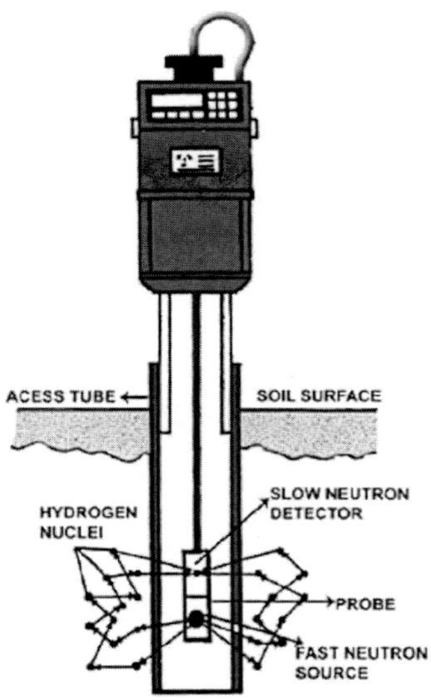

Fig. 7. Neutron probe

Merits	Demerits
More reliable	Costly
Very rapid method	Needs technical skill to operate the instrument
Soil conditions are not disturbed	Radiation hazard may affect the device as well as the operator. Hence periodical checkup is needed.
Repeated measurements are possible in the same locations	Small change in soil moisture content cannot be detected
Moisture content at different depths at smaller interval is possible in one stroke	Requires frequent calibration for different soil types

7.2.3 Soil moisture estimation by tensiometer

Tensiometers (Irrometers)are used to measure the soil moisture tension.It consists of a 7.5 cm long porcelain clay (porous ceramic) cup, a protective metalic tube, a vacuum gauge and a hollow metalic tube holding all parts together. At the time of installation, the system should be filled with water from the opening at the top and closed with rubber cork.When the tensiometer is installed in soil, water moves into and out of the ceramic cup in relation to the soil water tension. When water moves out of the cup into the soil, it creates a vacuum in the tube which is measured in the gauge. When desired tension is reached the field is irrigated.

Installation of a tensiometer

In this method, hollow pipe with a sharp edge, hammer, tensiometer and coloured stakes are required.

Procedure

Select the spot for installation. Bore a hole by driving a hollow pipe with sharp cutting edge in the soil by hammering it to the desired depth.Insert the tensiometer into the hole.Compact the soil around the stem of the tensiometer. Make a small heap so that water will not stagnate near the tensiometer. Take readings in the morning at 8.00 A.M. record the readings frequently so that the difference between two consecutive readings is not more than 10 centi-bars.Plot the readings on a paper against days. The following table gives the interpretation of the dial reading.

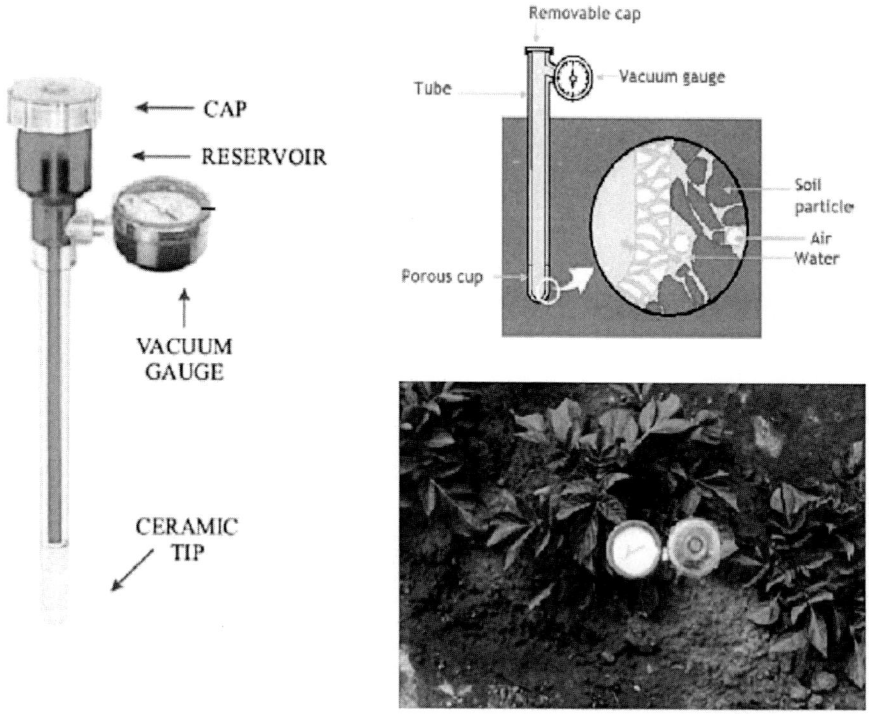

Fig. 8. Measurement of soil moisture tension by tensiometer

Table 4. Interpretation of tensiometer readings

	Dial reading in atmospheres	**Interpretation**
Nearly saturated	0 0.1	Near saturated soil often occurs for a day or two following irrigation. Danger of waterlogged soils, a high water table, poor soil aeration, or the tensiometer may have broken tension, if readings persist.
Field capacity	0.11 0.2 0.3	Irrigations discontinued in this range to prevent waste by deep percolation and leaching of nutrients below the root zone. Sandy soils will be at field capacity, in the lower range with clay soil at field capacity in the upper range.

Irrigation range	0.4	Usual range for starting irrigations. Soil aeration is assured in this range. In general, irrigations start at readings of 30-40 in sandy textured soils (loamy sands and sandy loams).
	0.5	Irrigations usually start from 40-50 on loamy soils (very fine sandy loam and silt loams).
	0.6	On clay soils (silty clay loams, silty clays, etc.) irrigations usually start from 50-60. Starting irrigations in this range ensures maintaining available soil moisture at all times.
Dry	0.7	This is the stress range. However, crop not necessarily damaged or yield reduced. Some soil moisture is readily available to the plant but is getting dangerously low for maximum production.
	0.8	Top range of accuracy of tensiometer, readings above this are possible but the tensiometer will break tension between0.8 and0.85 atmospheres.

(Stegman *et al.,* 1980)

7.2.4 Pressure plate and membrane

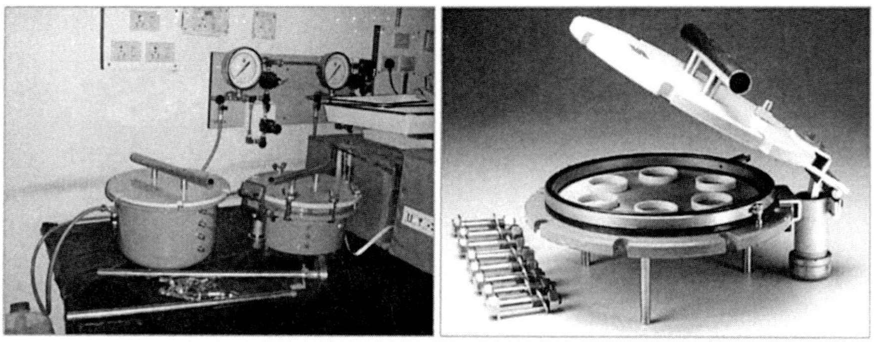

Fig. 9. Pressure plate and membrane apparatus

Laboratory measurements of soil water potential are usually made with pressure membrane and pressure plate apparatus. It consists of ceramic pressure plates or membranes ofhigh air entry values contained in airtight metallic chambers strong enough to withstand high pressure of 15 bars or more. The apparatus enables development of soil moisture characteristic curves over a wide range of matric potential.The porous plates are first saturated and then soil samples are placed on these plates. Soil samples are saturated with water and transferred to the metallic chambers. The chamber is closed with special wrenches to tighten the nuts and bolts with required torque for sealing it. Pressure is applied from a compressor and maintained at a desired level. It should be ensured that there is no leakage from the chamber. Water starts to flow out from saturated soil samples through outlet and continues to trickle till equilibrium against the applied pressure is achieved. Soil samples are taken out and oven dried to constant weight for determining moisture content on weight basis. Moisture content is determined against pressure values varying from −0.1 to −15 bars. The values of moisture content so obtained at a given applied pressure are used to construct soil moisture characteristic curves.

8. PHYSICAL CLASSIFICATION OF WATER

The water held within the soil pores is referred to as soil moisture. The manner in which it is held in the soil and to what extent it is translocated into plant system forms a basis of observation of drying of wet soils and of plants growing on these soils, water may be divided into three categories viz., gravitational water, capillary water and hygroscopic water (Fig.10).

8.1 Gravitational water

Water held between 0.0 to 0.33 bars (0 to −33 kPa) soil moisture tension, free and in excess of field capacity, which moves rapidly down towards the water table under the influence of gravity is termed as gravitational water. Even though, the gravitational water is retained with low energy, it is of little use to plants, because it is present in the soil for only a short period of time and while in the soil, it occupies the larger pores i.e., macro pores, thereby reducing soil aeration. Therefore, its removal from the soil profile through natural drainage is generally regarded as a pre-requisite for optimum plant growth and development.

8.2 Capillary water

As the name suggests capillary water is held in the pores of capillary size i.e., micro pores around the soil particles by adhesion (attraction of water molecules for soil particles), cohesion (attraction between water molecules) and surface tension phenomena. It includes available form of liquid water extracted by growing plants and is held between field capacity (0.33 bars or −33 kPa) and hygroscopic coefficient (31 bars or −3100 kPa). However, the water within the capillary range is not equally available i.e., it is readily available starting from 0.33 bars up to a certain point often referred to as critical soil moisture level (for most crops it varies between 20 to 50% depletion of available soil moisture) and thereafter up to 15 bars (−1500 kPa) it is slowly available. Further below, when the soil exerts tensions between 15 bars and 31 bars, the water is held very tightly in thin films and is practically not available for plant use. The capillary water moves in any direction, but always in the direction of increasing tension and decreasing potential.

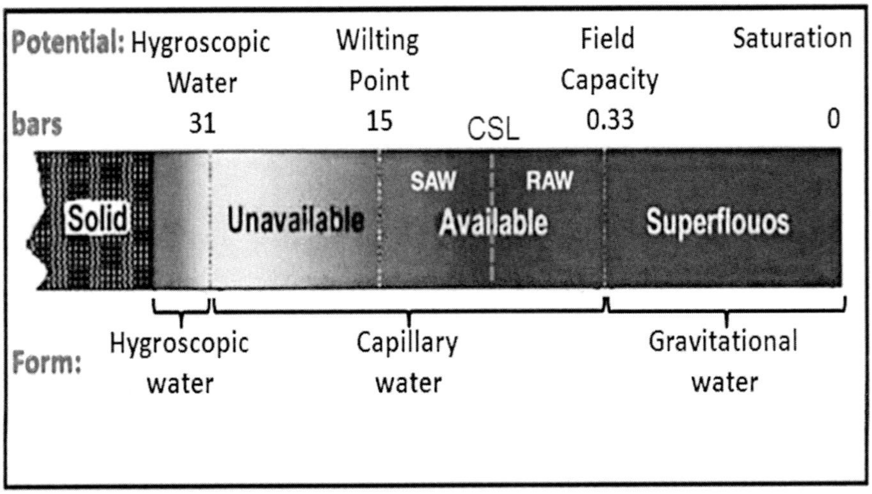

Fig. 10. Physical classification of soil water

8.3 Hygroscopic water

The water held tightly in thin films of 4-5 milli microns thickness on the surface of soil colloidal particles at 31 bars tension (−3100 kPa) and above is termed as hygroscopic water (Fig.). It is essentially non-liquid and moves primarily in vapour form. Plants cannot absorb such water because, it is held very tenaciously by the soil particles (i.e., >31 bars). However, some microorganisms may utilize it. Unlike capillary water which evaporates easily at atmospheric temperatures (i.e., it requires very little energy for its removal), hygroscopic water cannot be separated from the soil unless it is heated at 100°C and above for 24 hours.

8.3.1 Soil physical characteristics

Soil physical properties plays greater role in soil available water and irrigation management. The most important soil properties affect irrigation management are 1. Soil texture, 2. Soil structure, 3. Porosity and 4. Density of soil.

Soil texture

Soil texture is the most important and fundamental property of the soil that is most intimately related to soil water relationship. It refers to the relative proportion of mineral particles of various sizes in a given soil i.e., the proportions of coarse, medium and fine particles, which are termed sand, silt and clay, respectively. Various combinations of these fractions are used to classify soil according to its texture. Using the name of the predominant size fraction designates texture. Three broad and fundamental groups of soil textural classes are recognized as sand, loam and clay (Table).

Table 5. Common textural classes encountered in the field

Soil type	Texture	Basic soil textural class
Sandy soils	Coarse	Sands andloamy sand
Loamy soils	Moderately coarse	Sandy loam anfine sandy loam
	Medium	Very fine sandy loam, loam, silt loam andsilt
	Moderately fine	Sandy clay loam, silty clay loam andclay loam
Clayey soils	Fine	Sandy clay, Ssilty clay andclay

The textural class of a soil can be accurately determined in the laboratory by mechanical analyses. They are, Coarse sand: 2.0 – 0.2 mm; fine sand: 0.2 – 0.02 mm; silt: 0.02 – 0.002 mm; and clay: < 0.002 mm

The soil texture is closely related to:

(a) Water holding capacity of the soil
(b) Quantity of water to be given at each irrigation i.e., irrigation water depth
(c) Irrigation interval and number of irrigations
(d) Permeability i.e., ability of the soil to transmit water & air
(e) Infiltration rate

For example, coarse textured soils (sandy soils) have low water holding capacity and facilitate rapid drainage and air movement. Therefore, crops grown on these soils require frequent irrigations with less irrigation water depth at each irrigation. On the other hand, fine textured soils (clayey) have relatively high water holding capacity, however the permeability for water and air is slow thus, resulting in poor drainage and sometimes the soils get waterlogged. Considering its various effects, the soils with loam texture are the ideal soils for growing most crops under irrigated conditions.

Soil structure

The structure of a soil refers to the arrangement of the soil particles and the adhesion of smaller particles to form large ones or aggregates. On the surface, soil structure is associated with the tilth of the soil.

The dominant shape of aggregates in a horizon determines their structural type, such as speroidal (granular or crumbly subtypes), platy (plate and laminar subtypes); prism like (columnar or prismatic subtypes) and block like (cube and sub-angular subtypes). On surface, the soil structure is associated with tilth of the soil.

The soil structure influences primarily:

(a) Permeability for air and water
(b) Total porosity and in turn water storage capacity in a given volume of soil
(c) Root penetration and proliferation

Soils without definite structure may be single grain types, sands or massive types such as heavy clays. For example, a structure-less soil allows water to percolate either too rapidly or too slowly. Platy structure restricts the downward movement of water.

Crumbly, granular are most desirable for efficient irrigation water management and normal crop growth.

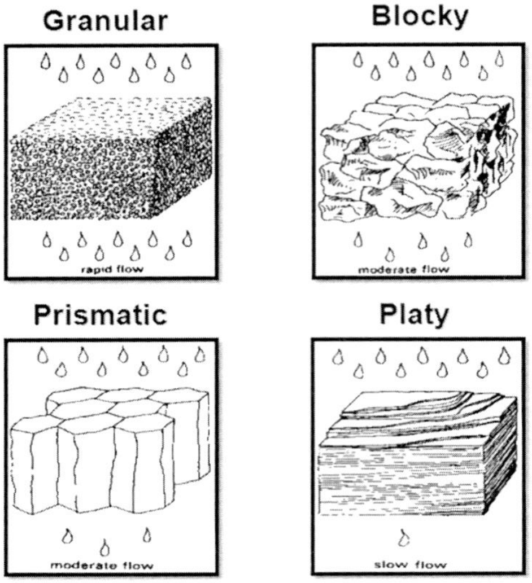

Fig. 11. Effect of structural type on permeability for water

Density of soil solids

There are two ways to express the soil weight – particle density and bulk density.

(a) Particle density (*rp*): It is the ratio of a given mass (or weight) of soil solids to that of its volume. It is usually expressed in terms of g/cm^3. Thus if 1 cm^3 of soil solids weigh 2.65g, the particle density measures 2.65 g/cm^3. Although considerable range may be observed in the density of individual soil minerals, the values for most mineral soils usually vary between the narrow limits of 2.6 to 2.75 g/cm^3. This soil property is independent of size of the soil particles and the arrangement of the soil solids. Knowledge of particle density is necessary for determination of total, capillary and non-capillary porosity.

(b) Bulk density (*rb*):It refers to the ratio of a given mass of an oven dried soil to that of its field volume (i.e. solids + pore spaces). The bulk density is calculated by the following formula:

$$\text{Bulk density (g/cm}^3) = \frac{\text{Weight of over dried soil (g)}}{\text{Total filed volume of soil (cm}^3)}$$

Knowledge of bulk density is of particular importance in the determination of depth of water for a given depth of soil and total, capillary and non-capillary porosity. The permeability of soil for water, air and penetration of plant roots is also influenced by bulk density. Compression or compaction of soil particles increases bulk density, consequently, it lowers the porosity in turn the soil water storage capacity. It directly affects the crop performance particularly where water availability is sub-optimal, which is a characteristic feature of light soils. The bulk density values for various soil textural classes are given in Table.

Table 6. Bulk density values of various soil types (USDA)

Soil texture	Bulk density (g/cm^3)	Soil texture	Bulk density (g/cm^3)
1. Sandy	1.60 – 1.70	6. Silty loam	1.40 – 1.50
2. Loamy sand	1.60 – 1.70	7. Silty clay loam	1.35 – 1.45
3. Sandy loam	1.55 – 1.65	8. Sandy clay loam	1.40 – 1.50
4. Fine sandy loam	1.50 – 1.60	9. Clay loam	1.30 – 1.50
5. Loamy soil	1.45 – 1.55	10. Clay soil	1.25 – 1.35

Porosity

Soil total porosity (f) is an index of pore volume in the soil. It is the space in a given volume of soil that is occupied by air and water or not occupied by the soil solids. The ratio of volume of voids to the total volume of soil is called as porosity.

The total porosity is calculated as follows:

Total porosity (f) (%) = 1 – (Bulk density / Particle density) × 100

Total porosity value generally lies in the range of 30 to 60% for arable soils. Coarse textured soils tend to be less porous (35–50%) than the fine textured soils (40–60%), though the mean size of individual pores is greater (>0.06mm in diameter) in the former than in the latter. Total porosity is inclusive of both, capillary (micro pores) and non-capillary porosity (i.e., macro pores). In a sandy soil, in spite of the relatively low total porosity, the movement of air and water is surprisingly rapid because of the dominance of the macro pores. Therefore the size of the individual pore spaces rather than their combined volume is an important consideration for optimum soil-water relations. For ideal conditions of aeration, permeability, drainage and water distribution, a soil should have about equal amount of macro and micro pore spaces. Knowledge of porosity in a given volume of soil is very important with respect to irrigation water management, because it is an index of moisture storage capacity and aeration conditions, the two most important factors that influence the plant growth and development.

9. PLANT WATER STRESS/MOISTURE STRESS

Soil moisture is the most limiting factor in farming situations. Understanding of the physiological processes that occur during moisture stress is necessary to ameliorate the stress effects either by management practices or by plant improvement. In biology, stress is the result of action of external factor on an organism. Moisture stress indicates the action of lack of or excess of water on plants. The term moisture stress is generally used for deficit moisture conditions though it is applicable to excess moisture also.

9.1 Development of moisture stress

Water deficits occur in the plant whenever transpiration exceeds absorption. It may be due to excessive water loss, reduced absorption of water from soil or both. After irrigation or rain, water deficits develop gradually and only for a very short period in the beginning. Through there is sufficient amount of soil moisture, water deficits develop due to higher transpiration than absorption especially on hot middays. This temporary wilting is known as incipient wilting or midday depression. As days progress, the soil moisture is depleted and plants cannot absorb sufficient amount of water to meet the high demand. Water deficits, therefore, occur for prolonged period. Under such conditions, stomata are open for a short period in the morning and evening and partially closed during the rest of the day. When the soil moisture reaches about –15 bars, plants show wilting symptoms most of the day, but do not die and it is known as wilting point. Plants recover when irrigation is given or rain is received. If the soil moisture is further allowed to deplete to a level of -60 bars, plants die permanently and this is known as ultimate wilting point.

9.2 Causes (plant responses) of moisture stress

Moisture stress does not affect all aspects of plant growth and development equally. Some processes are highly susceptible while others are far less affected. The final yield of the crop is the integrated result of these effects of stress on water relations, photosynthesis, respiration, nutrition, growth and development.

9.2.1 Water relations

Water deficits cause several changes in the plants. It alters the water status by its influence on absorption, translocation and transpiration. Moisture stressed plants manifest typical reduction in leaf water potential and relative water content. With the onset of dry conditions, a progressive and continuous decline in turgor of the plants is observed.

Water deficits also cause increase in leaf and canopy temperature. The stomata gradually close down reducing transpiration and its cooling effect on leaves. The increase in leaf temperature may be up to 8°C over ambient temperature.

9.2.2 Photosynthesis

Photosynthesis is reduced by moisture stress due to reduction in photosynthetic rate, chlorophyll content, leaf area and increase in assimilate saturation in leaves.

The photosynthetic process i.e. entry of carbon dioxide into the leaf, photochemical reactions and reduction of carbon dioxide are affected by moisture stress. The gaseous exchange through leaves is governed by boundary layer (thin layer of still air on the leaf surface), stomatal and mesophyll resistances. As moisture stress increases, stomata start closing as a mechanism to resistances. As moisture stress increases, stomata start closing as a mechanism to reduce tranpirational losses. As a consequence, the entry of carbon dioxide is also reduced. The conductance of mesophyll cells gradually decreases with the age of the leaf. However, the decrease in conductance of mesophyll cells occurs earlier under moisture stress. The decreases in mesophyll conductance of carbon dioxide also effects photosynthesis. However, decrease in photosynthetic rate is mainly due to stomotal resistance.

Leaf Area: Reduction in photosynthesis due to moisture stress in mainly by the reduction in leaf area than by photosynthetic rate. Decreasing the tissue water potential by only –1 bar or less results in a perceptible decrease in cellular growth while photosynthetic rate is affected only at –16 bars. The reduction in functional leaf area may be due to breakdown of chlorophyll, reduction of leaf expansion, tillering or branching and increase in leaf senescence in moisture stressed plants. The elongation rates of expanding leaves are more reduced due to moisture stress than net photosynthesis.

Assimilate Saturation: Translocation of assimilates is affected by water stress. Due to assimilate saturation, photosynthesis is reduced. Between the two processes viz. Translocation of photosynthates and photosynthesis, translocation is more reduced due to moisture stress than net photosynthesis.

9.2.3 Respiration

Respiration increases with mild stress. However, as the water deficits become severe, it decreases. In wheat carbon dioxide output is more in early stages of drought before there is any measurable change in water content. More severe drought lowers water content and respiration.

9.2.4 Anatomical changes

Periodical water stress leads to anatomical changes like decrease in size of the cells and intercellular spaces, thicker cell wall and greater development of mechanical tissue. The number of stomata is fixed for known number of epidermal cell, but sizes of cells are reduced without reduction in the number of stomata per unit number of cells. Therefore, stomata per unit leaf area tend to increase under moisture stress.

9.2.5 Metabolic reaction

Almost all metabolic reactions are affected by water deficits. Severe water deficits cause decrease in enzymatic activity. However, these effects are selective. The amount of enzymes involved in hydrolysis* remain or increase resulting in breakdown of starch and protein while peroxidase activity decreases. Accumulation of sugars and amino acids takes place under moisture stress. Proline, an amino acid, accumulates whenever there is moisture stress. Its accumulation is more in later stages of plants and it is considered as a good Indicator of moisture stress.

9.2.6 Hormonal relationships

As a consequence of water deficits, hormonal balnce is altered. The activity of growth promoting hormones like cytokinin, gibberellic acid and indoleacetic acid decreases and growth regulating hormone like abscisic acid, ethylene, betain etc., increases. The translocation of growth promoting hormones is also reduced by moisture stress. Abscisic acid content is inversely related to leaf water potential. Abscisic acid acts as water deficit sensor to minimize the loss of tissue water potential. It controls stomata and thus reduces the water loss when there is moisture stress. Ethylene production inducted by moisture stress is considered to be the cause for leaf and fruit drop. Betain is another hormone produced by the moisture stressed plants and it is used as an indicator of moisture stress.

9.2.7 Nutrition

Moisture stress affects fixation, uptake and assimilation of nitrogen. Nitrogen fixation by leguminous plants is reduced by moisture stress due to reduction in leghaemoglobin in nodules, specific nodules activity and number of nodules. There is negative correlation between specific nodule activity and stomatol resistance. Nitrogenase activity in the nodules, which indicates the specific nodule activity, is less in stressed plants compared to those which are irrigated. In addition, the dry weights of nodules are significantly less in moisture stressed plants. Moisture stress also delays nodule formation in leguminous crops, but number of bacteroids containing cells are not affected.

Nutrient uptake is the product of nutrient content and dry matter produced. Moisture stress may or may not reduce nutrient content but reduces dry matter production considerably. As a result, nutrient uptake is reduced. There is considerable amount of evidence to show that N, P and K uptake is reduced by moisture stress. N P K uptake and transpiration rate are highly correlated even under mild moisture stress conditions. Nitrogen assimilation is also affected by moisture stress mainly due to the reduction in nitrate reductase activity. Nitrogen deficiency increases stomatal resistance. Phosphorus deficiency increases the sensitivity of the most recently expanded leaves to moisture stress and stomata close while there is still sufficient turgor in mesophyll cell.

9.2.8 Growth

Generally, the organ growing most rapidly at the time of stress is the one most affected. The expansion of cells and cell division are reduced due to moisture stress resulting in decrease in growth of leaves, stems and fruits. Moisture stress affects germination, leaf area, leaf expansion and root development.

9.2.9 Development

In general, moisture stress delays maturity. However the developmental response of plants to moisture stress depends on stage of occurrence of stress. The general rule is that if stress occurs before flowering, the duration of the crop increases and when it occurs after flowering, the duration decreases.

9.2.10 Reproduction and grain growth

Moisture regime during flowering and grain development determines the number of fruits and individuals grain weight respectively. During ripening which involves dehydration and certain biochemical processes, moisture regime has very little effect on yield components. For many crop plants, especially cereals, moisture stress at panicle initiation is critical. As the panicle is the organ that is growing most rapidly, it is most affected by stress due to reduction in cell expansion. Anthesis is another important moisture sensitive stage as moisture stress at this stage causes drying of pollen and loss of viability of pollen. Stress during grain development also reduces yield due to reduction in leaf area and photosynthesis. However, vegetative and grain filling stages are less sensitive to moisture stress.

9.2.11 Yield

The effect of water stress on yield depends largely on what proportion of the total dry matter produced is considered as useful material to be harvested when the yield consists of mostly aerial parts like forage crops, tobacco etc., the effect of moisture stress is the same as those on total growth. When the yield consists of underground storage organs as in potatoes, sugar beet etc., it will be as sensitive as total growth. When the yield is fibre or chemicals where economic product is a small fraction of total dry matter, moderate stress on growth does not have adverse effect on yields.

9.3 Crop adaptations

The ability of crop of grow satisfactorily under water stress is called drought adaptation. Adaptation is structural or functional modification in plants to survive and reproduce in a particular environment.

Crops survive and grow under moisture stress conditions mainly by two ways: (I) escaping drought and (II) drought resistance

Flow chart showing different mechanism for overcoming moisture stress

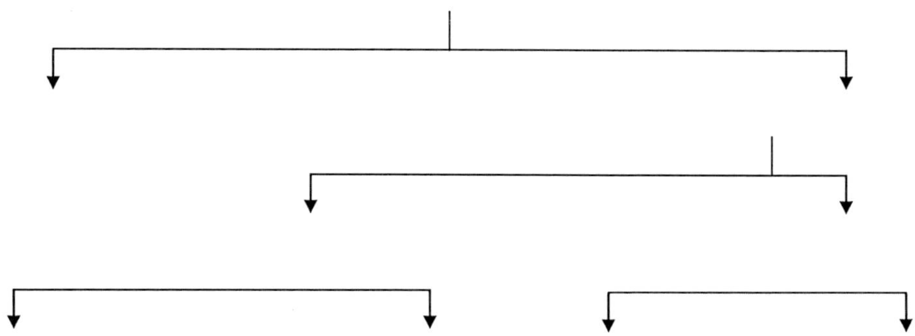

9.3.1 Escaping drought

Evading the period of drought is the simplest means of adaptation of plants to dry conditions. Many desert plants, the so called ephemerals, germinate at the beginning of the rainy season and have an extremely short life period (5 to 6 weeks) which is confined to the rainy period.

In cultivated crops, the ability of a cultivar to mature before the soil dries is the main adaptation to growth in dry regions. Certain varieties of pearl millet mature within 60 days after sowing. Short duration pulses like cowpea, green gram, black gram can be included in this category.

9.3.2 Drought resistance

Plants can adopt to drought either by avoiding stress or by tolerating stress due to different mechanisms. These mechanisms provide drought resistance, but they may or may not reduce productivity.

9.3.2.1 Avoiding stress

Stress avoidance is the ability to maintain a favourable water balance, and turgidity even when exposed to drought conditions, thereby avoiding stress and its consequences. A favourable water balance under drought conditions can be achieved either by (i) conserving water by restricting transpiration before or as soon as stress is experienced; or (ii) accelerating water uptake sufficiently so as to replenish the lost water.

9.3.2.1.1 Mechanisms to conserve water

Stomatal mechanism: Drought resistant varieties are capable of regulating the stomatal mechanism and closing stomata when drought prevails. Drought resistant varieties open their stomata more rapidly in the early morning when moisture stress is at its minimum and photosynthesis can proceed with the least loss of water.

Increased photosynthetic efficiency: When stomata are closed as a mechanism against water loss, photosynthesis is affected due to reduction in entry of carbon-di-oxide through stomata. Some plants overcome this problem by their increased photosynthetic efficiency. In C_4 Plants, the carboxylating enzyme namely phosphoenlo-pyruvic acid carboxylase has very high affinity for carbon dioxide and high potential activity. Therefore, for the same amount of stomatal opening, C_4 plants have higher photosynthetic rate than C_3 plants. Thus, C_4 plants are said to be drought resistant as they are able to grow better even under moisture stress. The C_4 plants like sorghum, maize, pearl millet, and sugarcane can photosynthesize even when carbon dioxide concentration is 100ppm.The plants with crassulacean acid metabolism (CAM) are highly drought resistant. They open stomata only during night when they take carbon dioxide into the leaves and store it as organic acid. During the day, as the stomata are closed, carbon dioxide does not enter the leaves.

Lipid deposits on leaves: Some plants like soybean, sorghum etc., reduce water loss by depositing lipids on plant surfaces under moisture stress.

Reduction in leaf area: Most of the plants reduce transpiration by limiting leaf area of the plants. The total leaf area of plants is a product of number of branches or tillers, number of leaves per branch and individual leaf area. Water deficits reduce tillering or branching which in turn reduces the leaf area resulting in less transpiration. The individual leaf size is reduced as leaf expansion is very sensitive to moisture stress.In grasses, the leaves roll or curl due to moisture stress and thus reduce the area exposed to solar radiation resulting in low transpiration. Leguminous plants show parahelionastic movements *i.e.,* the leaves are oriented parallel to sun rays thus avoiding the load of solar radiation.Senescence or drooping of leaves is another mechanism for reducing leaf area. Senescence of lower leaves which are not contributing any photosynthesis to grain yield may be advantageous but premature senescence of leaves causes reduction in yield.

Leaf surface: various morphological characteristics of leaves help to reduce the transpiration rate and may influence survival of plants under drought conditions. Leaves with thick cuticle, waxy surface and spines etc., are common and effective. Pubescence increases leaf reflectance and reduces solar radiation incidence.

Effect of awns: Awned varieties give more yield under drought conditions compared to awnless varieties. Awns contribute 12 per cent of photosynthates to grain. Compared to transpiring area awns, it s contribution of photosynthates to grain is more.

Water storage in plants: water stored in plants is considered as an adaptive mechanism. However, its significance is small in crop plants because their total water content is frequently less than the water they transpire in a day.

9.3.2.1.2 Mechanism to improve water uptake

Drought avoidance is promoted by well developed deep root system with high efficiency to extract water from deeper layers of soil. This mechanism is desirable only if there is sufficient soil moisture in deeper layers for extraction. Water uptake can be improved by several mechanisms.

Efficient root system: Deep, well branched and rapidly growing root system helps in absorbing more moisture by exploiting higher volume of soil. It is an important morphological adaptation that helps in drought resistance without losing productivity.

Root-shoot ratio: If the roots are more compared to transpiring shoot, water balance can be maintained. Drought increases root growth and root-shoot ration which is an important mechanism of drought avoidance.

Increase in liquid phase conductance: To maintain high water potential in plants, not only uptake is important, but also conductance. Lowering of resistance to water can be achieved by increasing either diameter of xylem vessels or their number.

Osmotic adjustment: Stress induced break down of carbohydrates and proteins increase the concentration of solutes in the cell sap resulting in reduction of osmotic potential. Plants which have the ability to lower their osmotic potential in response to slowly developing deficits are useful. Osmotic adjustment increases translocation and helps in increasing grain yield. Abscisic acid produced in the leaves at lower turgor reduces seed set through its effect on pollen viability.

9.3.2.2 Drought tolerance

Due to different drought avoidance mechanisms, plants are able to maintain favourable water balance and adverse effects of reduced water potential are not felt by the plants. In drought tolerance, water potential of plant is reduced and its adverse effects are felt. Drought tolerance can be defined as tolerance of the plants to a level of stress at which 50 per cent of cells die.

9.3.2.2.1 Mitigating stress

The simplest way of mitigating stress is by resisting dehydration and by maintenance of higher osmotic pressure by accumulating higher amounts of solutes. The leaves with thick cuticle resist cell collapse.

9.3.2.2.2 High degree of tolerance

When plants are subjected to severe moisture stress, stomata are closed permanently resulting in cessation of photosynthesis and starvation of plants. However, plants capable of keeping the stomata partially open can photosynthesize and survive. Young leaves, are more resistant to drought than older leaves due to higher protein content. Plastic strain indicates irrevocable loss of plant tissues due to severe stress. The mechanism involved is killing of transpiring portions and keeping propagation material without desiccation. In many species of perennial plants, the above ground parts die off during drought and underground parts such as rhizomes, bulbs, tubers etc., remain alive but dormant.

Developmental mechanisms

Adaptations in phenological development has been demonstrated in some crops. Mild water deficits at late vegetative stage hastened ear emergence and maturity. Phenological plasticity is available in plants with indeterminacy, branching and tillering. Small amount of indeterminacy, branching and tillering is essential for flexibility for varying ecological conditions. In crop plants, greatest advances in breeding for water limited environments have been achieved by shortening life cycle.

Morphological adaptations

Genotypes which exhibit changes in leaf angle and recover quickly after stress are drought resistant. Most of the legumes and sunflower show these foliar movements.

Anatomical adaptations

In order to maintain high water potential in the leaves, particularly under high evaporative demand, plants must be able to absorb higher amount of water and transport it quickly to leaves.

Physiological adaptations

Plants which close their stomata at high transpiration demands and open quickly as transpiration demand decreases are helpful. The ability of plants to continue at a relatively high rate of photosynthesis during moisture stress is one of the criteria for selection of a drought resistant variety. Drought induced accumulation of proline is another criteria in certain crops like barley, groundnut etc., which are less sensitive to moisture stress at critical periods. Survival of seedlings and adult plants can be used as a measure of drought tolerance. Plant survival is an important criteria where periodic severe drought occurs during early and midseason. Crops with C4 and CAM type of photosynthesis are preferred in dry regions. Selection of C_3 plants with high biochemical efficiency for fixing carbon dioxide at less concentration is useful.

Remobilisaiton of reserves

There are two sources of assimilate supply for grain development of cereals and legumes *i.e.,* from photosynthesis before anthesis and after anthesis. Under normal conditions, contribution of assimilates to grain is less than 20 percent in most plants except in rice where it ranges from 20 to 40 per cent. But under moisture stress conditions, it may be up to 50 to 75 percent. Limited success in identifying wheat or barley lines which transfer more pre anthesis assimilates to the grain under stress than under non-stress conditions should not deter others from seeking this character in a wide range of genotypes.

Miscellaneous characteristics

Different plant characteristics like horizontal orientation of leaf, large peduncle, large flag leaf, more number of grains, bold grains and awns are considered as desirable characters in cereals.In brief, the desirable plant characteristics for drought resistant crop are:

1. Rapid germination and early establishment of deep roots.
2. Rapid phonological development.
3. Developmental plasticity
4. Parahelonastic movements
5. Stomata sensitive only to large vapour pressure deficits and insensitive to low leaf water potential.
6. Ability to adjust osmotically.
7. Large transfer of assimilates from stem to grain, and
8. Dehydration tolerance, particularly at seedling and grain filling stages.

In addition to the above, characteristics like efficient root system, capacity of roots to grow in dry soils and high rate of photosynthesis with thick small leaves are also desirable for drought resistance.

9.4. Methods to overcome plant water stress

(a) Nutritional Management: Among the major nutrients, potassium and magnesium are found to be highly deficient due to water deficit conditions. Therefore, application of potassium enhances the water uptake as well as the water relations in the plant tissues by osmo regulation processes, by acting as a potent osmo regulator (osmolyte), thereby the solute potential is reduced. Besides, potassium nutrition also helps in the favourable stomatal regulatory mechanisms, which regulate the water balance of the plants. This has also resulted in the increased WUE of the plants. Similarly, magnesium is component of chlorophyll, its content and uptake is drastically reduced due to the water stress effect. This is most prominent in Mg-loving crops like cotton. Therefore, foliar application of the following nutrients depending upon the occurrence of their efficiencies will mitigate the water-stress induced nutritional imbalance in crops.

(a) 2 % DAP

(b) 0.5 to 1 % potassium chloride (KCl)

(c) 0.5 % Zinc sulphate

(d) 0.5 – 1.0 % Ferrous sulphate + 1 % urea

(e) 0.3 % Boric acid

(b) Use of Antitranspirants: Chemical substances used to reduce the transpiration is called as antitranspirant

Classification of Antitranspirants: Four types

I. Materials causing stomatal closure

1. Herbicides like 2, 4 – D, Phosphon D and Atrazine
2. Fungicides like Phenyl Mercuric Acetate (PMA)
3. Metabolic inhibitors like hydroxy sulfonates, potassium metabisulphite *etc.*
4. Growth hormones like ABA, Ethrel, TIBA, succinic acid, ascorbic acid and Cycocel (CCC)

II. Reflectant Types

1. Kaoline
2. China Clay
3. Calcium bicarbonate
4. Lime water

III. Thin-forming chemicals

1. Hexadecanol (Higher alcohols)
2. Cetyl alcohol
3. Methanol
4. Paclobutrazol
5. Brassinolide
6. Resorcinol

IV. Polyethylene materials forming thick films

1. Mobileaf
2. Folicot
3. Waxol
4. S- 800
5. Hico-110R

(c) Use of plant growth regulators (PGR)

Cycocel & Mepiquat chloride: For promoting root growth (for more water absorption) and suppressing leaf area development (for reducing transpiration loss of water) and delaying on set of leaf senescence.

Cytokinins and Salicylic acid: They delay the leaf senescence processes and also favour stem reserve utilization by the developing grains especially during the water deficit situations.

Brassinolides: These PGRs increase the photosynthetic activity of the plants

Ascorbic acid: Ascorbic acid acts as an anti-oxidant agent for scavenging Reactive Oxygen Species (ROS) accumulating under stress and thus avoiding membrane damage.

(d) Pre-sowing Hardening of Seeds / Plants: Hardening of seeds / plants to required temperature / chemicals enables the plants to overcome the specific stresses. This process actually hardens the protoplasm (by osmo regulation), which enables the seeds to absorb more water under favourable situations to maintain its viability under unfavourable conditions.

Chemicals used for seed hardening process especially under rainfed conditions

1. 1% KCl
2. 1% KH_2PO_4
3. 100 ppm Succinic acid
4. 0.5% NaCl
5. 100 ppm $ZnSO_4$
6. 100 $MnSO_4$
7. 100 ppm Ascorbic acid
8. 250 ppm Cycocel
9. 0.5% $MgSO_4$

QUESTIONS

I. Fill in the blanks

1. Soil is a _______ phase system
2. Solid portion of the soil made up of_______, ______ and ________.
3. Soil contains more ________ and less _________.

4. The water contents expressed under certain standard conditions are referred as ____________.
5. Expand SPAC: __________
6. Tensiometers (irrometers) are used to measure the ___________.
7 Spirit burning method is not recommended for soils with high ______content.

II. Choose the best answers

1. Proportion of solid, liquid and gases in the soil is _________.
   (a) 50 : 25 : 25 (b) 50 : 20 : 25 (c) 50 : 50 : 25 (d) 25:50:25
2. _________ is the force of attraction between molecules of different substances
   (a) Adhesion (b) Cohesion (c) Adsorption (d) None of these
3. _________is the force of attraction between molecules of same substances
   (a) Adhesion (b) Cohesion (c) Adsorption (d) None of these
4. _________ capacity refers to the condition of soil at which all the macro and micro pores are filled with water.
   (a) Available soil moisture (b) Saturation
   (c) Wilting point (d) None of these
5. Chemicals used for seed hardening process
   (a) KH_2PO_4 (b) Cycocel (c) A & B (d) None of these

III. Answer the following questions

1. What is soil moisture constant?
2. What is the difference between adhesion and Cohesion?
3. Soil moisture tension
4. Saturation capacity
5. Field capacity
6. Permanent wilting point
7. Available soil moisture
8. What is the effective root zone depth?
9. What is hygroscopic coefficient?
10. Write the importance of soil moisture constants in irrigation management
11. Soil water movement
12. Infiltration
13. Seepage
14. Permeability
15. Percolation
16. Types of water movement
17. Write three importance of water plant relationship

18. Effective root zone depth
19. Soil-Plant-Atmosphere Continuum (SPAC)
20. Differences between evaporation and transpiration
21. Factors affecting ET
22. Importance of evapotranspiration (ET) and potential evapotranspiration (PET) for crop plants
23. Potential evapotranspiration (PET)
24. Explain different mechanism for overcoming moisture stress
25. Drought adaptation.
26. Mechanisms to conserve water in plants
27. Mechanism to improve water uptake in crops
28. Methods to overcome plant water stress
29. What is Anti-transpirants and classify?
30. Explain the soil water movement

3

CROP WATER REQUIREMENT-CRITICAL STAGES FOR IRRIGATION-WATER REQUIREMENT FOR DIFFERENT FIELD CROPS-WATER USE EFFICIENCY

1. CROP WATER REQUIREMENT

Different losses like percolation, seepage, runoff etc., occur during transport and application of irrigation water. Water is needed for special operations such as land preparation, transplantation, leaching etc. Water requirement of a crop (*WR*), therefore, includes evapotranspiration (*ET*), application losses and water needed for special purpose.

$$WR = ET + \text{Application losses} + \text{Water for special purposes}$$

Water requirement is a demand; whereas, the supply consists of contribution from irrigation water, effective rainfall (*ER*) and soil profile contribution including that from shallow water table (*S*).

$$WR = IR + ER \text{ (Effective rainfall)} + S \text{ (Soil profile contribution)}$$

2. IRRIGATION REQUIREMENT

Irrigation requirement is the total amount of water applied to a field to supplement rainfall and soil profile contribution to meet the water needs of crops for optimum growth.

$$\text{Irrigation requirement } (IR) = WR - (ER + S)$$

The net irrigation requirement is the amount of irrigation water just required to bring the soil moisture content in the root zone depth of the crops to field capacity. Thus, the net irrigation requirement is the difference between field capacity and soil moisture content in the root zone before irrigation. Gross irrigation requirement is the total of net irrigation requirement and other losses such as conveyance, distribution and application or total depth of water required for entire crop period.

3. COMPONENTS OF CROP WATER REQUIREMENT

3.1 Evaporation

Evaporation is defined as the process by which water moves out of the water or soil surface in the form of water vapour to atmosphere due to solar energy (heat energy). Evaporation from natural surface such as open water or bare soil is a diffusive process by which water in the form of vapour is transferred from the underlying surface to the atmosphere.

3.2 Transpiration

Transpiration is the process by which water in plant body transfers to the atmosphere in the form of water vapour.Transpiration is the process by which water evaporates in the form of water vapour from living plant body especially from leaves to atmosphere.

3.3 Evapotranspiration (ET) or Consumptive Use (CU)

Evaporation and transpiration from an unit area planted with or covered with crop is known as evapotranspiration (*ET*).

Evapotranspiration is almost equal to Consumptive Use (*CU*)

Consumptive use includes losses due to evapotranspiration and water used by the plants for its metabolic activities.

$$CU = E + T + \text{Water needed for metabolic purpose}$$

Since the water used in the actual metabolic processes is insignificant (less than 1 % of *ET*), consumptive use is generally taken equivalent to *ET*.

Hence, $ET = CU$

3.3.1 Classification of consumptive use of water by crops

Consumptive use of water by crops may be estimated for any specified period. It may be classified as daily consumptive use, seasonal consumptive use and peak period consumptive use based on the specific period of cropping seasons.

3.3.1.1 Daily consumptive use

The amount of water consumptively used during 24hour period is called the daily consumptive use. It is estimated usually to record the peak period use rates of water by crops for formulating the cropping pattern and to decide the water supply from sources during different periods of cropping seasons.

3.3.1.2 Seasonal consumptive use

The total amount of water used in evapotranspiration by a cropped area during the entire growing season is called as seasonal consumptive use. It is expressed as depth of water in cm. This value is used to evaluate and determine the seasonal irrigation water requirement.

3.3.1.3 Peak period consumptive use

The highest consumptive use for certain period in the growing season is called as peak period consumptive use. It is used in planning and designing of the irrigation system. The peak period consumptive use varies with soil, crop and climatic condition.

3.4 Application losses

Different losses like percolation, seepage, runoff etc., occur during transport and application of irrigation water.

3.5 Special need

Water is needed for special operations such as land preparation, transplantation, leaching etc.

4. FACTORS AFFECTING WATER REQUIREMENT

The water requirement of any crop is dependent upon

(1) **Crop factors:** Variety, growth stage, duration, plant population and growing season

(2) **Soil factors:** Texture, structure, depth and topography

(3) **Climate factors:** Temperature, relative humidity and wind velocity

(4) **Crop management practices:** Tillage, fertilization and weeding etc.

Water requirement of crops vary from area to area and even from a field to a field on a farm depending on the above mentioned factors. It is, therefore, not appropriate to make general statements of water requirement of a crop or specify the number of irrigation or irrigation interval for a crop. Instead, indicating the development stages of crop when it is essential to give irrigations and avoid moisture stress is more meaningful. The need for irrigation at other stages of crop growth are best decided based on the amount of moisture available in the soil and needs of the crop.

4.1 Crop factors

Varieties of the same crop differ in duration, rooting pattern and canopy structure. The variety, with longer duration obviously requires more water for completion of the life cycle. During the growth of crop, consumptive use is maximum during flowering and grain filling in cereals compared to that in seedling stage. Crops differ in producing leaf area and covering the ground. Higher the leaf area index, more is the evapotranspiration. Evapotranspiration also differs with height of the crop. Tall crops intercept more solar radiation and have more evapotranspiration than short crops.

4.2 Soil factors

Evapotranspiration from soils differ due to their difference in hydraulic conductivity, reflectivity and thermal conductivity. At higher moisture regimes, coarse textured soils have higher hydraulic conductivity than fine textured soils. With the result, evaporation is faster in coarse textured soils under intermittent wetting and drying. Evaporation mostly occurs from the top 5 cm of soil and soil structure up to 15 cm depth influences evaporation through its influence on water supply to evaporation site. Higher percentage of aggregate of more than 1.0 mm diameter, reduce the upward movement of water and hence, evaporation. Formation of ridges and furrows reduces evaporation due to the presence of large sized aggregates. Colour of the soil also has considerable influence on evaporation from the soil surface. Dark coloured soils absorb more of solar radiation and thus, increase evaporation.

4.3 Climate factors

It is well known that evapotranspiration is strongly influenced by solar radiation, temperature, relative humidity and wind velocity. In addition, advective energy also influences evaporation. Hot and dry area surrounding the irrigation are increases evaporation. Advection is a serious problem in arid and semi-arid regions.

4.4 Crop management practices

All the management practices that provide favourable environment to the crop increase the leaf area and thus, increase evapotranspiration. Tillage practices influence rooting characteristics and thus influence evapotranspiration. Frequent irrigation results in higher proportion of evaporation. Weeding reduces competition for moisture and increases irrigation interval. Similarly, mulching reduces evapotranspiration considerably.

5. METHODS OF DETERMINATION OF WATER REQUIREMENT

5.1 Transpiration Ratio method

The amount of water transpired by a crop to produce a unit of dry matter is called transpiration ratio. In earlier days, the values of transpiration ratio was used to determine the water needs of plants. This method of estimating water needs of plants has several drawbacks. Transpiration ratio is measured in green house on potted plants which is considerably different from field grown plants. The knowledge of transpiration ratio is, therefore, of little value in planning and irrigation under field conditions.

5.2 Depth-interval-yield method

Field experiments are conducted with different depths of irrigation water for each irrigation and various intervals between irrigations. The total water used for obtaining maximum yield is considered as water requirement of the crop. The disadvantage of this method is variation in climate in different seasons which has considerable influence on water requirement of crops.

5.3 Water balance method

Evapotranspiration from a large area like a watershed can be estimated by water balance method. It is the estimation of incoming and outgoing water from an area. Water is added to the soil by precipitation (P), irrigation (I) and groundwater contribution (S). The different losses were evaporation (E), transpiration (T), deep percolation (D) and surface runoff (R).

5.4 Field experiments

Crops are grown with best management practices and with best known irrigation schedule indifferent seasons. Water applied, effective rainfall and soil contribution are measured. Water needed during the entire crop season is the water requirement of the crop.

The water requirement for different seasons are worked by this method. This is accurate method, but, it is time consuming, laborious and costly.

5.5 Climatological methods

The main component of water requirement is evapotranspiration. Evapotranspiration mainly depends on climate and it can be estimated with the help of empirical formulae. Once the potential evapotranspiration is estimated, crop evapotranspiration is obtained by multiplying potential evapotranspiration with crop coefficients. Water requirements of crop is obtained by adding application losses and water needed for special operations.

5.6 Drum culture technique for rice

Different components of irrigation water in rice field are percolation, evaporation and transpiration and ineffective rainfall, can be measured by the drum culture technique. Four metallic square drums of size 50 × 50 cm and 120 cm height, two with bottoms and two without bottoms are embedded in the rice field leaving 20 cm above the soil surface. In the tanks *A* and *B* with bottoms, the excavated soil is replaced layer wise to simulate the soil conditions outside the drum. The tanks *C* and *D* without bottoms are embedded in soil, without disturbing the inner core by excavating the soil only outside the tank.

Rice seedlings are transplanted in drums *A*, *C* and *D* on the same day as the surrounding plot. In drum *B*, rice seedlings are planted on the same day after removing the roots. These rootless plants are replaced every week with plants of the same age and size grown as a bulk crop and with roots cut off. These plants without roots in drum *B* shade the soil as plants in the other drums shade the soil. Plant population, fertilizers and other cultivation operations are same. Both inside and outside the drums. Water level is maintained at 5 cm depth or desired in the drums and in the surrounding plot. Water level in the drums is measured with point gauge daily at 8.00 A.M. and losses due to *ET* and *p* ercolation are replenished to maintain 5 cm continuous submergence or any replenished different components of water losses are calculated as follows.

$$\text{Evapotranspiration } (ET) = \text{Loss in tank } A = E + T$$

$$\text{Evaporation } (E) = \text{Loss in tank } B$$

$$\text{Percolation } (P) = C - A$$

$$\text{Transpiration } (T) = A - B$$

Drum *D* is fitted with 0.5 cm diameter tubes at one cm interval for precise water control and also to let out ineffective rainfall. The outlet tube of any required depth is connected to a porcelin jar embedded in the ground outside the drum to collect ineffective rainfall. The mouth of a jar is closed to avoid direct rainfall but a small vent is provided to allow air to escape and facilitate collection of ineffective rainfall into the jar.

Ineffective rainfall = Excess water from drum *D* collected in the jar after rainfall

Effective rainfall = Daily rainfall – ineffective rainfall

Effective rainfall (*ER*) = Total rainfall – Total Ineffective rainfall

Table 7. Water requirement of different crops

Crop	Water requirement (mm)	Crop	Water requirement (mm)
Rice	900-2500	Tomato	600-800
Wheat	450-650	Potato	500-700
Sorghum	450-650	Pea	350-500
Maize	500-800	Onion	350-550
Sugarcane	1500-2500	Bean	300-500
Sugar beet	550-750	Cabbage	380-500
Groundnut	500-700	Banana	1200-2200
Cotton	700-1300	Citrus	900-1200
Soybean	450-700	Grape	500-1200
Tobacco	400-600	Pineapple	700-1000

6. WATER USE EFFICIENCY

Irrigation plays a vital role in Indian agriculture. The impact of irrigated agriculture on water resources is significant since it uses 85% of the total water resources in the country. The limits to the availability of water and land for irrigated agriculture necessitate the careful use of water resources for achieving higher water use efficiency.

The term 'water use efficiency' originates in the economic concept of productivity. Productivity measures the amount of any given resource that must be expended to produce one unit of any good or service. In a similar manner, water use efficiency measures the quantity of water taken up by the crop during its crop life to produce a unit quantity of the output i.e., crop yield. In general, lower the water resource input requirement per unit of crop yield produced, the higher the efficiency.

Further water use efficiency is closely related to water conservation. The growing water scarcity and the misuse of available water resources are nowadays major threats to sustainable agricultural development. Therefore, water use efficiency has a clear role to play in sustainable development, in other words, the use of the earth's water resources by today's inhabitants while assuring that future generations have sufficient capacity to meet their own needs.

6.1 Water use efficiency (WUE)

It is expressed in kg/ha/cm. The proportion of water delivered and beneficially used on the project can be calculated using the following formula

$$Eu = \frac{Wu}{Wd} \times 100$$

Where, Eu = water use efficiency (%)

Wu = water beneficially used

Wd = water delivered

Water use efficiency is also classified as (i) crop water use efficiency and (ii) field water use efficiency.

6.1.1 Crop water use efficiency (CWUE)

It is otherwise called consumptive water use efficiency. It is the ratio of crop yield (Y) to the sum of the amount of water taken up and used for crop growth (G), evaporated directly from the soil surface (E) and transpired through foliage (T) or consumptive use (CU).

$$\text{CWUE} = Y/(G + E + T)$$

Where,

$$(G + E + T) = CU$$

In other words, ET is CU since water used for crop growth is negligible.

$$\text{CWUE} = Y/CU$$

It is expressed in kg/ha/mm or kg/ha/cm.

6.1.2 Field water use efficiency (FWUE)

It is the ratio of yield of crop (Y) to the total amount of water used in the field.

$$\text{FWUE} = Y/WR$$

Where,

FWUE - Field water use efficiency

WR - Water requirement

This is the ratio of crop yield to the amount of water used in the field (WR) including growth (G), direct evaporation from the soil surface (E), transpiration (T) and deep percolation loss (D).

$$\text{FWUE} = Y / (G + E + T + D)$$

$$G + E + T + D = WR$$

It is expressed in kg/ha/mm (or) kg/ha/cm

Of the two, the crop water use efficiency is more of research value whereas the field water use efficiency has grater practical importance for planners and farmers.

6.2 Methods to improve WUE

The WUE can be improved by improving genetic and agronomic practices in addition to the change of climatic characters.Factors affecting water use efficiency is grouped under three major criteria as follows

Genetic factors

(i) Crop species

(ii) Crop varieties

Climatic factors

(i) Temperature
(ii) Precipitation
(iii) Sun light
(iv) Wind velocity
(v) Relative humidity

Agronomic factors

(i) Time of sowing
(ii) Depth of sowing
(iii) Direction of sowing
(iv) Plant population
(v) Fertilization
(vi) Insect pests and diseases
(vii) Irrigation method
(viii) ET control measures

Genetic factors

Water use efficiency of C_4 plant species such as maize, sorghum, sugarcane, pearl millet, finger miller etc. is higher when compared to C_3 species such as pulses, oilseed crops, wheat, barley, oats, etc.

Crop varieties also differ in WUE. High yielding varieties, hybrids, GM crops, etc. due to their dwarf plant type, responsive to water & fertilizer, pest & disease resistance and high harvest index exhibit higher WUE as compared to traditional varieties characterized by rank vegetative growth, low harvest index, susceptible to lodging, pests & diseases.

Climatic factors

The amount of solar radiation determines the rate of photosynthesis and hence, the potential yield. Other components of climate *viz.,* temperature, day length, rainfall, etc. influence vital physiological processes and thereby determine the actual harvested yield. The lower the relative humidity is, the greater will be the ETc. Therefore, low relative humidity in the atmosphere increases transpiration without any corresponding increase in dry matter production and will reduce WUE. Light and temperature that normally affect both transpiration and dry matter production will either increase or decrease WUE according to which of the two predominates. High wind velocity increase ETc without any concurrent increase in dry matter production hence, decrease WUE.

Crop management factors

Time of sowing: Timely sowing ensures optimal temperatures, soil moisture availability and other soil physical conditions favouring optimal crop growth and development with greater ability to compete with prevalent weed flora, hence increases WUE.

Depth of sowing: Optimal depth of sowing improves seedling emergence, vigour and finally crop yield, hence improves WUE.

Direction of sowing: North south row orientation of crop rows influences the interception and utilization of incident solar radiation which in turn influences crop yield and improves WUE as compared to east west direction of row pattern.

Plant population: Optimal plant population promotes uniform & rapid development of crop canopy without any competition for growth resources *viz.,* light, nutrients, water, CO_2 etc. hence improves WUE.

Fertilization: Fertilization of crops suffering from low nutrition under adequate soil water availability decreases crop yield considerably, with a relatively small increase in crop evapotranspiration, therefore, markedly improves WUE.

Insect pests and diseases: Insect pests and diseases reduce crop yield as well as WUE to varying degrees depending upon the intensity of infestation, because ETc or water requirement of crop will not change to a significant level except in cases where premature death of plants occurs.

Irrigation method: Field water use efficiency in general is higher with over head sprinkler, micro-sprinkler and drip methods of irrigation as compared to surface irrigation methods *viz.,* furrow, border strip, check basin etc. owing to higher crop yield and lower seasonal water application.

ET control measures: Use of mulches, anti-transpirants, shelterbelts and elimination of weeds etc reduce water losses from cropped field in terms of soil evaporation and transpiration without any reduction in crop yield, hence, markedly improve WUE.

Table 8. Crop water use efficiency under various irrigation methods

Crop	Methods of irrigation	Water requirement (cm)	Water saving (%)	Yield (kg/ha)	Yield increase (%)	Water use efficiency (kg/ha mm)
Banana	Drip	97.00	45.00	87500	52.00	90.20
	Surface	176.00	0	57500	–	32.67
Sugar cane	Drip	94.00	56.00	170000	33.00	180.85
	Furrow	215.00	–	128000	–	59.53
Aerobic rice	Conventional aerobic rice	74.30	38.10	4747	–	6.39
	Surface drip	61.90	48.40	5940	14.20	9.60
	Sub-surface drip	61.90	48.40	6227	19.80	9.74
	Conventional transplanted	120.00	–	5200	–	4.33

Cotton	Drip	28.00	66.27	3250	25.00	116.10
	Furrow	83.00	0	2600	–	31.33

7. CRITICAL STAGES OF WATER REQUIREMENTS

The crop plants in their life cycle pass through various phases of growth, some of which are critical for water supply. The most critical stage of crop growth is the one at which a high degree of water stress would cause maximum loss in yield or irrecoverable yield loss. It is also known as moisture sensitive period. Further, studies on irrigation at growth stages may give an indication as to whether scarce water can be used more efficiently by scheduling irrigation's at critical stages. Scheduling of irrigations on the basis of critical growth stages is simple and easy for the farmers. However, it does not take into account the available soil water in the crop root zone depth. Excessive irrigations without significant soil/ plant water deficit could be harmful to crop plants and might reduce their yield under certain situations. The criterion may not hold well in long duration crops like sugarcane, cotton; crops requiring frequent irrigation's viz., potato or standing/nearly standing water (rice) and where there is interference by rainfall of different amounts.

Table 9. Critical stages of different crop for irrigation

Crop	Critical stages / Sensitive stages
Cereals and millets	
Rice	Panicle initiation, Heading and flowering
Sorghum	Flowering and grain formation (dough)
Maize	Tasseling, silking and dough stage
Cumbu	Heading and flowering
Ragi	Primordial initiation and flowering
Wheat	Crown root initiation, tillering and booting (CRI and Flowering)
Oilseeds	
Groundnut	Flowering, peg formation and pod development
Sesame	Blooming to maturity
Sunflower	Two weeks before and after flowering
Soybean	Blooming and seed formation
Safflower	From rosette to flowering
Castor	Full growing period
Pulses	Pre flowering and pod development
Cash crop	
Cotton	Flowering and boll formation
Sugarcane	Formative phase
Tobacco	Immediately after transplanting

Vegetables	
Onion	Bulb formation to maturity
Tomato	Flowering and fruit setting
Chillies	Flowering
Cabbage	Head formation to maturity
Legumes	
Alfalfa	Immediately after cutting for hay crop and flowering for seed crop
Beans	Flowering and pod setting
Peas	Flowering and pod formation
Others	
Coconut	Nursery stage root enlargement
Potato	Tuber initiation and maturity
Banana	Throughout the growth
Citrus	Flowering, fruit setting and enlargement
Mango	Flowering
Coffee	Flowering and fruit development

At critical stages, favourable water level should be ensured through timely irrigations

QUESTIONS

I. Fill in the blanks

1. _______ includes evapotranspiration (ET), application losses and water needed for special purpose.
2. ______is the total amount of water applied to a field to supplement rainfall and soil profile contribution.
3. Evapotranspiration is almost equal to ______.
4. The amount of water transpired by a crop to produce a unit of dry matter is called ____________.
5. __________________ method used to determine the water requirement of Rice crop.

II. Choose the best answers

1. Water requirements is _________.
   (a) Irrigation requirement (b) Effective rainfall
   (c) Soil profile contribution (d) All the above
2. Irrigation requirement is _____________.
   (a) WR (b) $WR - ER$ (c) $WR - (ER + S)$ (d) None of these

3. Consumptive use includes losses due to __________ & _________.
   (a) ET (b) Metabolic activities
   (c) Evaporation (d) ET + Metabolic activities
4. The water used in metabolic processes is less than ______ % of ET
   (a) 5 % (b) 4 % (c) 1 % (d) 22 %
5. Evapotranspiration from a large area like a watershed can be estimated by __________ method
   (a) Transpiration ratio (b) Field experiment
   (c) Climatological (d) Water balance

III. Answer the following questions

1. Difference between crop water requirement and irrigation requirement
2. Components of crop water requirement
3. Classification of consumptive use of water by crops
4. Factors affecting water requirement
5. Methods of determination of water requirement
6. Transpiration ratio method
7. Drum culture technique for rice
8. Water use efficiency
9. Crop water use efficiency
10. Field water use efficiency
11. Methods to improve water use efficiency
12. Moisture sensitive period
13. Define: Irrigation requirement, Net irrigation requirement and Gross irrigation requirement

4

METHODS OF IRRIGATION - MERITS, DEMERITS, SUITABILITY OF SURFACE AND SUB-SURFACE IRRIGATION METHODS

Criteria for selection of irrigation method

- Water supply source
- Topography
- Quantity of water to be applied
- Crop / variety
- Method of cultivation

1. IRRIGATION METHODS

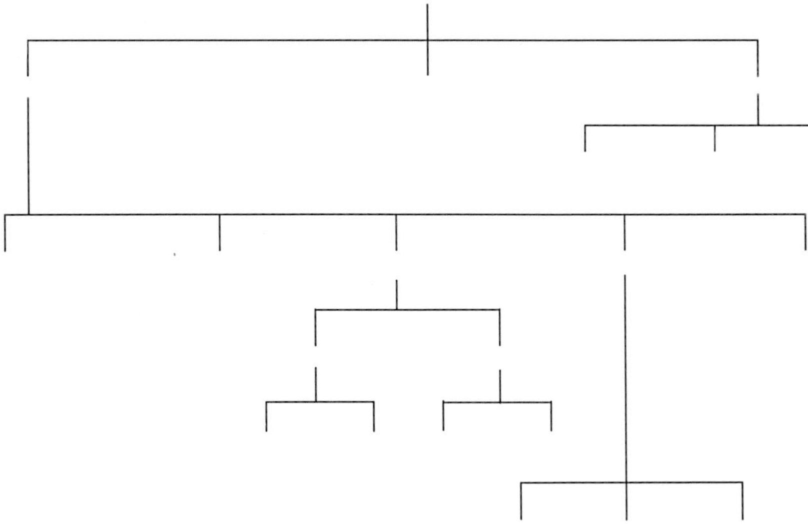

2. SURFACE IRRIGATION

Oldest (4000 years back) and most common method. 90% of world's total irrigated area is under this method. In USA also, 66% is by surface irrigation. This method is most suitable for low to moderate infiltration rates and leveled lands and having <2-3% slope. It is labour intensive. Different methods are given below.

2.1 Flooding

It consists of applying water to the field without any bunds to guide the flow of water to wet the soil surface completely. Generally, it is practiced only when irrigation water is abundant and where land leveling is not followed. Sometimes, it is also adopted in the initial stages of land development. This method is most commonly used for irrigation of crops sown by broadcasting method viz., rice, low value pastures, lawns and millets etc.

Advantages

- No land leveling & land shaping
- Low labour and land preparation costs
- Less skill required by irrigator

Disadvantages

- Applied water is lost by deep percolation & surface runoff
- Low irrigation application efficiency

2.2 Border strip irrigation

- The land is divided into number of long parallel strips called borders.
- These borders are separated by low ridges.
- The border strip has a uniform gentle slope in the direction of irrigation.
- Each strip is irrigated independently by turning the water in the upper end.
- The water spreads and flows down the strip in a sheet confined by the border ridges.

Suitability: To soils having moderately low to moderately high infiltration rates. It is not used in coarse sandy soils that have very high infiltration rates. Suitable to irrigate all close growing crops like wheat, barley, fodder crops and legumes and not suitable for rice.

Advantages

1. Border ridges can be constructed with simple farm implements like bullock drawn "A" frame ridger or bund former.
2. Labour requirement in irrigation is reduced as compared to conventional check basin method.
3. Uniform distribution of water and high water application efficiencies are possible.
4. Large irrigation streams can be efficiently used.

Width of border strip: It varies from 3-15 rn

Border length

Slope	Soil	Length
0.25 - 0.60%	Sandy and sandy loam	60-120 m
0.20 - 0.40%	Medium loam soil	100-180 m
0.05 - 0.20%	Clay loam and clay soil	150-300 m

2.3 Beds and channel or Check basin irrigation

- It is the most common method.
- Here the field is divided into smaller unit areas so that each has a nearly level surface.
- Bunds or ridges are constructed around the area forming basins within which the irrigation water can be controlled.
- The water applied to a desired depth can be retained until it infiltrates into the soil.
- The size of the basin varies from 10m to 25 rn depending upon soil type, topography, stream size and crop.

Adaptability

- Small gentle and uniform land slopes.
- Soils having moderate to slow infiltration rates.
- Adapted to grain and fodder crops in heavy soils.

Advantages

1. Check basins are useful when leaching is required to remove salts from the soil profile.
2. Rainfall can be conserved and soil erosion is reduced by retaining large part of rain.
3. High water application and distribution efficiency.

Limitations

1. More area occupied by ridges and field channels.
2. Precise land grading and shaping are required.
3. Labour requirement is higher.
4. Not suitable for crops which are sensitive to wet soil conditions around the stem.

2.4 Basin method (Ring basin)

This method is a modification of check basin method and is suitable for sparsely grown orchard crops and cucurbits. In this method, a circular or square bund is constructed around each tree/plant or group of plants/trees to create a basin for irrigation. These basins are suitably connected to irrigation conveyance channels is such a way that either each basin is irrigated separately or group of basins by flowing water from one basin to another through inter-connections.

Advantages

- High irrigation application efficiency can be achieved with properly designed system.
- Unskilled labour can be used, as there is no danger of erosion.

Disadvantages

- High labour requirement
- Bunds restrict use of modern machinery in the field
- Limited to relatively uniform lands

2.5 Ridges and furrow irrigation

- Used in the irrigation of row crops.
- The furrows are formed between crop rows.
- The dimension of furrows depend on the crop grown, equipment used and soil type.
- Water is applied by small running streams in furrows between the crop rows.
- Water infiltrates into soil and spreads laterally to wet the area between the furrows.

Adaptability

1. Wide spaced row crops including vegetables.
2. Suitable for maize, sorghum, sugarcane, cotton, tobacco, groundnut, potatoes.
3. Suitable to all the soils

Advantages

I. Water in furrows contacts only one fifth of the land surface.
2. Labour requirement for land preparation and irrigation is reduced.
3. Compared to check basins there is less wastage of land in field ditches.

2.5.1 Types of furrow irrigation

(i) Based on alignment of furrows:

1. Straight furrows
2. Contour furrows

(ii) Based on size and spacing

1. Deep furrows
2. Corrugations

1. Deep furrow method: In furrow method of irrigation, the flat bed surface is converted into a series of ridges and furrows running down the slope. The spacing of the furrow is ordinarily determined by the spacing of row crop. The length of the furrow and slope depends on several factors *viz.,*texture, intake rate etc. Depending on the soil, crop spacing, farm equipment used etc. spacing between furrows varies from 60 to 120 cm. Depending on the soil texture furrow length ranges from 20 to 300 m or even more. To avoid overtopping and scouring problems, furrow inflow rates are normally limited to 2 to 15 m^3/hr/furrow. This method can be used either with small or large streams of irrigation water because it can be diverted into any number of furrows. Slope along the furrows may range from 0.2 to 2%. Where the land is too sloppy (>5%), the furrows must be constructed on contours.

Among the various surface irrigation methods, there is a relative saving of water in furrow method; hence efficient use of water is possible. This method is commonly used for irrigating crops like potato, sugarcane, maize, cotton, melons, sugarbeets and vegetables like lettuce.

When furrow irrigation is practiced under saline and alkaline conditions, the lateral movement of soil moisture coupled with evaporation causes salt to accumulate in the ridges between furrows. If the salt accumulation reaches harmful levels, planting is advocated on the relatively salt free bottom of the furrows following pre-plant irrigation.

Advantages

- Fairly high irrigation application efficiency among surface irrigation methods.
- Furrows serve as field drains in areas of heavy rainfall.
- Low evaporation losses.

Disadvantages

- Not suitable in coarse textured soils with high infiltration rates.
- Possibility of intra-furrow soil erosion.
- Labour intensive.

2. Corrugation: This is a special method of furrow irrigation. Corrugations or rills are shallow furrows running down the slope from head ditches or laterals, which are sometimes very close to each other. Water moves down through several corrugations simultaneously and soaks laterally through the soil, wetting the area between the corrugations. This method is used for irrigating close growing crops, which do not require inter-culture operations and may also be used in conjunction with border irrigation. Corrugations are often used in fine textured soils that take water slowly and in soils, which tend to seal over and crust when flooded. The spacing and size of the corrugations vary with soil type, but in general, the more porous the soil, the more closely spaced the corrugations to permit rapid wetting between them without excessive deep percolation losses. The length of the corrugations should be such that upper end of the field has not been over irrigated by the time the lower end has received sufficient water.

(iii) Based on irrigation

- **All furrow irrigation:** Water is applied evenly in all the furrows and are called furrow system or uniform furrow system.
- **Alternate furrow irrigation:** It is not an irrigation layout, but, a technique for water saving. Water is applied in alternate furrows. For eg. During first irrigation if the even numbers of furrows are irrigated, during next irrigation, the odd number of furrows will be irrigated.
- **Skip furrow irrigation:** They are normally adopted during the period of water scarcity and to accommodate intercrops. In the skip furrow irrigation, a set of furrows are completely skipped out from irrigation permanently. The skipped furrow will be utilised for raising intercrop. The system ensures water saving of 30-35%. By this method, the available water is economically used without much field reduction.

2.6 Broad bed furrow (BBF)

Here beds of 1.5 m width, 15 cm height and convenient length are formed, separated by furrows of 30 cm width and 15 cm depth. Crops are sown on the beds at required intervals. It is suitable for heavy black soils and deep red soils. The furrows have a gradient of 0.6%. Broad bed furrow has many advantages over other methods.

- It can accommodate a wide range of crop geometry i.e close as well as wide row spacing.
- It is suitable for both sole cropping and intercropping systems.
- Furrows serve to safely guide runoff water in the early part of rainy season and store rain water in the later stages.
- Sowing can be done with seed drills.
- It can be formed by bullock drawn or tractor drawn implements. Bed former cum seed drill enables BBF formation and sowing simultaneously, thus reducing the delay between rainfall receipt and sowing.

2.7 Surge irrigation

Surge irrigation is the application of water in to the furrows intermittently in a series of relatively short ON and OFF times of irrigation cycle. It has been found that intermittent application of water reduces the infiltration rate over surges thereby the water front advances quickly. Hence, reduced net irrigation water requirement. This also results in more uniform soil moisture distribution and storage in the crop root zone compared to continuous flow. The irrigation efficiency is in between 85 and 90%.

3. SUB-SURFACE IRRIGATION

- In subsurface irrigation, water is applied beneath the ground by creating and maintaining an artificial water table at some depth, usually 30-75 cm below the ground surface.

- Moisture moves upwards towards the land surface through capillary action
- Water is applied through underground field trenches laid 15-30 rn apart.
- Open ditches are preferred because they are relatively cheaper and suitable to all types of soil.
- The irrigation water should be of good quality to prevent soil salinity.

Advantages

- Minimum water requirement for raising crops
- Minimum evaporation losses
- No wastage of land
- No interference to movement of farm machinery
- Cultivation operations can be carried out without concern for the irrigation period.

Disadvantages

- Requires a special combination of natural conditions.
- There is danger of water logging
- Possibility of choking of the pipes laid underground.
- High cost.

4. MICRO IRRIGATION

4.1 Drip irrigation or trickle irrigation

Drip irrigation is defined as the precise, slow and frequent application of small quantities of water to the soil in the form of discrete drops, continuous drops, and tiny streams through emitters located at selected points along a water delivery lateral line.

4.2 Sprinkler irrigation

Sprinkler Irrigation refers to application of water to crops in the form of spray from above the crop like rain. It is also called the overhead irrigation as water is allowed to fall as spray from above the crop. Sprinkler irrigation also called as pressurized irrigation as water under pressure is carried and sprayed into the air above the crop through a system of pipe network.

4.3 Rain gun irrigation

Rain gun irrigation is a type of sprinkler irrigation and covers larger area in a short time and suitable for all the crops particularity for high statured crops like sugarcane, tapioca, cotton and orchards.

QUESTIONS

I. Fill in the blanks

1. ___________ is the oldest and most common method of irrigation
2. Field is divided into smaller unit areas in ______________ method of irrigation
3. Corrugation method of irrigation suitable for ____________ crops
4. Water is applied beneath the ground by creating and maintaining an artificial water table is known as _____________

II. Choose the best answers

1. Which method is most commonly used for irrigation of crops sown by broadcasting
   (a) Border strip irrigation (b) Drip irrigation
   (c) Furrow irrigation (d) Flood irrigation
2. ___________ is a modification of check basin method and is suitable for sparsely grown orchard crops
   (a) Border strip irrigation (b) Furrow irrigation
   (c) Ring basin irrigation (d)None of the above
3. Ridges and furrow irrigation suitable for
   (a) Broadcasted crop (b) Row crop
   (c) Close spaced crops (d) Orchards
4. Applicationof water in to the furrows intermittently in a series of relatively short ON and OFF times of irrigation cycle is
   (a) Surge irrigation (b) Check basin
   (c) Ridges & Furrow (d) None of the above
5. Drip irrigation otherwise called as ___________ irrigation
   (a) Sprinkler (b) Rain gun
   (c) Trickle (d) None of the above

III. Answer the following questions

1. List out the criteria used selection of irrigation method
2. Draw the flow chart of various irrigation methods
3. Explain the BBF method of irrigation?
4. Advantages and disadvantages of sub surface irrigation methods
5. Give details on type of furrow irrigation
7. Advantage and disadvantage of surface irrigation
8. What is corrugation?
9. What is surge irrigation?
10. What is drip irrigation ? and write advantages and disadvantages
11. What is sprinkler Irrigation? and write advantages and disadvantages

5

MICRO IRRIGATION SYSTEM (DRIP & SPRINKLER IRRIGATION) - GOVERNMENT SUPPORT TO POPULARIZE MICRO IRRIGATION SYSTEMS

1. MICRO IRRIGATION

Micro-irrigation covers large number of irrigation practices whose common characteristics are the relatively small cross sections of the supply and distribution lines, the low water emission rates per distributary as well as localized delivery of water to a limited area. The important micro-irrigation systems are drip, micro-sprinkler, micro-jet systems and bubbler irrigation (suitable for tree crops and horticultural crops and lawns).

2. DRIP IRRIGATION OR TRICKLE IRRIGATION

Drip irrigation is defined as the precise, slow and frequent application of small quantities of water to the soil in the form of discrete drops, continuous drops, and tiny streams through emitters located at selected points along a water delivery lateral line.It differs from sprinkler irrigation by the fact that only part of the soil surface is wetted.

Current drip irrigation technology dates back to the work of Symcha Blass (1964). Based on the observation that a large tree near a leaking faucet exhibited a more vigorous growth than other trees in the area, he developed and first patented drip irrigation system. From Israel, the concept spread to Australia, North America and South Africa by the late 1960s, and eventually throughout the world. The availability of low cost plastic pipe for water delivery lateral lines helped to speed up the field use of drip irrigation system. At present, the largest area is in the USA and Spain (1.5 million ha each) followed by India (1.43 million ha). Total coverage in the world has increased from 4000 ha in 1972 to over 8 million ha in 2008.

2.1.1 Surface drip irrigation

The application of water to the soil surface as drops or tiny streams through emitters with discharges rate for point-source emitters less than 8 l/hour for single outlet emitter and for line-source emitters less than 4 l/hour. Often, the terms drip and trickle irrigation are considered synonymous.

2.1.2 Subsurface drip irrigation

The application of water below the soil surface through emitters, with discharge rate generally in the range of 0.6 to 3 l/hour.

Advantages

- Water saving (25-50%) losses due to deep percolation, surface runoff and transmission are avoided. Evaporation losses occuring in sprinkler irrigation do not occur in drip irrigation.
- Fertilisers can be applied through irrigation water (drip fertigation)
- Increases the yield of crops – 20-100%.
- Uniform water distribution.
- Application rates can be adjusted by using different size of drippers.
- Suitable for wide spaced row crops, particularly coconut and other horticultural tree crops.
- Soil erosion is reduced.
- Better weed control.
- Land saving and less labour cost.

Suitable slopes

- Drip irrigation is adaptable to any farmable slope.
- Normally the crop would be planted along contour lines and the water supply pipes (laterals) would be laid along the contour also.
- This is done to minimize changes in emitter discharge as a result of land elevation changes.

Suitable soils

- Drip irrigation is suitable for most soils. On clay soils water must be applied slowly to avoid surface water bonding and runoff.
- On sandy soils higher emitter discharge rates will be needed to ensure adequate lateral wetting of the soil.

Suitable irrigation water

- One of the main problems with drip irrigation is blockage of the emitters.
- All emitters have very small waterways ranging from 0.2-2.0 mm in diameter and these can become blocked if the water is not clean.
- Drip irrigation is particularly suitable for water of poor quality (saline water).

2.2 Components of drip irrigation

The components of drip irrigation system can be broadly grouped under head control unit, water distribution unit and delivery unit the details of which is elaborated hereunder.

2.2.1 Head control unit

At the head of the field is a set of appliances that make up the control head of any micro irrigation system. The function of control head for drip system is to organize, control and to regulate uniform and constant flow of water and nutrients to the field. It includes pump, filtering unit and essentially fertilizer injection equipment with accessories viz., pressure regulator, pressure gauge and water measuring devices. This is the most vital part of the system and proper care should be taken in its planning, operation and maintenance.

(i) Pumping unit: The pressure necessary to force water through the components of the system is obtained by a pump of suitable capacity. Centrifugal pumps in open wells and submersible pumps in bore wells are most suitable to provide sufficient pressure for operation the system. For small area, over head tank having a height of about 10 to 15 m can be used.

(ii) Valves: There are many types of valves used in the drip irrigation system.

- Bye-pass valve
- Control valve
- Air/vacuum valve
- Non-return valve

Bye-pass Valve

Sometimes when water from pump comes with high pressure, this valve can be used. This valve diverts some water again to the well. This problem can be solved to some extent with the gate valve itself. If the pressure is too high, Bye-pass valve can be used. If this problem does not exist, this valve is not needed.

Gate (Control) valves

Gate valves are widely used. They are used for allowing or shutting off water along any pipeline. There are three types of gate valves.

1. Ball Valve
2. GI Valve
3. Brass Valve

Ball valve is normally made of plastic. GI valve is made of Galvanized Iron. Brass valve is costly.

Air Valve

When pump is shut off, after water from pipes drained out, vacuum gets formed in pipes. To fill this vacuum, air gets into the pipe through drippers. During this process, soil particles are also sucked in through the drippers. Therefore, if an Air valve is provided for each sub main, this problem can be reduced to a larger extent. This valve is to be provided at the location of the highest elevation of sub main or main.

Non-return valve

Non-return valves allow the flow to pass through only one direction. When flow directions are reversed, the non-return valve closes. This valve is very important when fertigation and chemical application through irrigation water is done. Otherwise back flow may cause polluting the water source.

(iii) Fertilizer injection equipment: Fertilizer injection equipment is to be provided at the head of the drip irrigation system for applying fertilizers in solution directly to the field along with the irrigation water. The equipment for fertigation is available in different types and models differing in their properties. Three main groups of equipment, with different types and models in each group are available.

(a) Venturi injector: This operates on the venture principle, where a flow constriction with specific entrance and exit conditions in the pipeline creates vacuum because of the increased velocity of flow through the constriction. This creates a suction effect which is sufficient to suck fertilizer solution from an open container into the water flow. The rate of flow is depending on the pressure difference created. It has a control valve in main line and a control valve in bye-pass line. The bye-pass is connected with a venturi. At the throat section, a lateral pipe is connected.

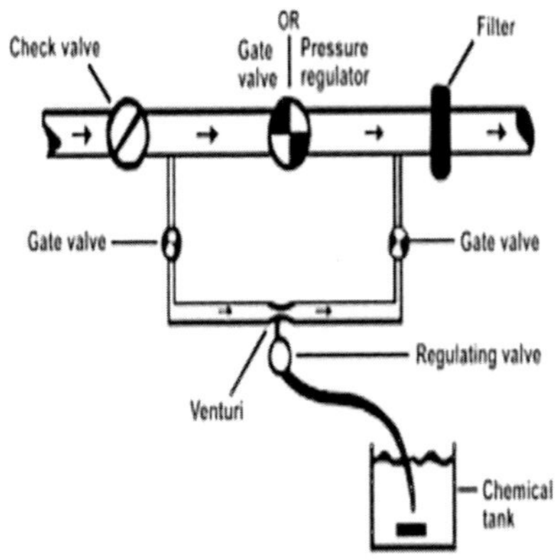

That pipe is dipped into fertilizer solution. While operation, the control valve in main line is partially closed or fully closed. The control valve in bye-pass is opened. When water passes through the throat section, suction pressureiscreated. Therefore fertilizer solution is sucked in. This is a simple and inexpensive method of fertilizer application. But it had some disadvantages. The pressure loss across a venture valve is high, about 1/3 of the operating pressure and precise regulation of flow is difficult because the rate of injection is very sensitive to the pressure and the rate of flow in the system.

(b) Pressure differential by – pass tank: This method uses a fertilizer in which dry or liquid fertilizer is used. The tank is connected to the main line by means of a bye-pass, so that some of the irrigation water flows through the tank and dilutes the fertilizer solution. The bye-pass is brought about by a control valve in the main line, which creates pressure gradient between the entry and exit of the fertilizer tank.

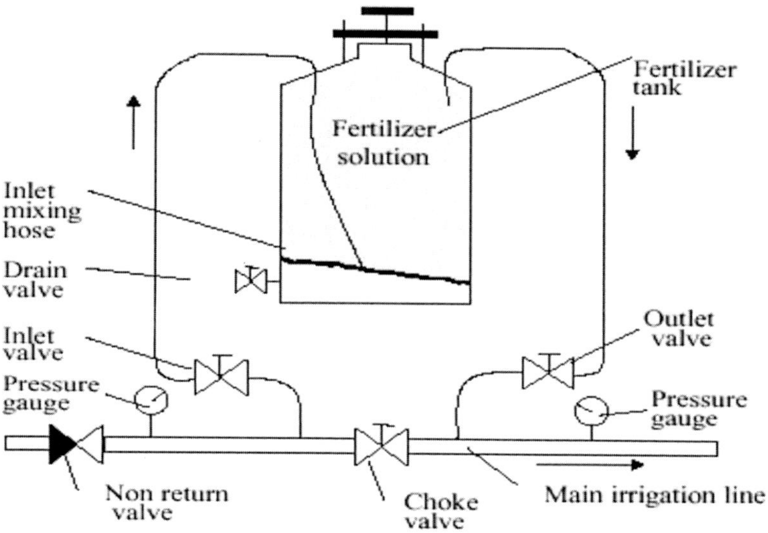

(c) Fertigation pumps: In direct injection pump, pumps of piston type or diaphragm pumps are most commonly used due to their reliability and accuracy of application pumps can be adjusted over a scope of different ranges to provide a continuous and uniform concentration of nutrients in the irrigation water. There is no pressure loss in the system. Injection devices allow central fertilization; the whole process may be controlled from one central point.

(iv) Filtration unit: Removal of suspended particles is very essential to get optimum performance of the drip system, since irrigation water is rarely free of suspended material. The particle size, which can be tolerated in the system, depends on the emitter construction. Typically, the recommendation is for the removal of particles larger than one tenth the diameter of the orifice or the flow passage of the emitter. One reason for this is that several particles may group together and bridge the passage way. This is typical with organic particles having about the same density of water. Another reason that heavier inorganic such as fine and very fine sands tends to settle and deposit at inside walls of emitters. The type of filter needed depends on the quality of water and operating pressure of drip system. Filters are to be installed at the head of the system and four types of filter, sand filter, screen filter, hydro cyclone filter and disc filter are most commonly used.

(a) Sand filter: Sand or gravel or media filters are used as primary filters, when water used for irrigation is received from river, pond, reservoir, open well etc. They are used principally for filtering out heavy loads of suspended impurities like sand and silt particles and algae and other inorganic materials. They are cylindrical tanks having layers of coarse, sand and fine gravel of selected sizes with a provision of valves for flushing the filter assembly in case of clogging. The sizes are ranging from 500 - 900 mm diameter.

(b) **Screen filter:** Screen filters are the simplest and least expensive filters. If water contain fewer impurities, screen filters alone will be sufficient to remove the impurities. Otherwise screen filters are fitted in series with gravel filter to remove solid impurities, ie they are used as a secondary filter. Hence, they are used as primary filter when water is received from bore well and as secondary filter, when is water received from river, pond, open well, reservoir etc. Screen filter consists of a single or double perforated cylinders placed in plastic or metallic container for removing the impurities. Generally 100 to 200 mesh screens are used in this type of filters. The inlet provides tangential entry of water to the casing so that heavier particles are thrown out wards and only small particles move towards the strainer at the centre. The tangential entry also helps in reducing frictional losses. It must be cleaned and inspected periodically for satisfactory operation. If head loss is higher than the permissible value, which is provided by the manufacturers, it needs cleaning. If no pressure change occurs for a long period, it indicated that the screen or seals are broken or mesh size too large, which necessitates replacement.

(c) **Hydro cyclone filter:** If irrigation water is containing large amount of sand, hydro-cyclone type filters are required. It is also known as 'vortex sand separator' or centrifugal sand separator. They depend on centrifugal force to remove and eject high- density particles from the water. The inlet provides tangential entry into the filter body so that heavier particles are thrown on the periphery of conical body of the filter and finally rolled down to under flow chamber. The collected sand and other material may be emptied manually or by using special flushing valve.

(d) **Disk Filter:** Disk filters are efficient in filtration of organic material and algae. They facilitate high volume filtration. The filter elements consists of a stack of discs, which held together by longitudinal plastic rods, capture debris. During filtration, disks are pressed together. There is an angle in the alignment of two adjacent disks, resulting in cavities of varying size. Disk filters are available in wide range from 400 – 600 mesh. Back flushing can be done manually by stopping the water flow, opening the disc housing and directing water jet to start spinning motion.

2.2.2 Water distribution unit

The distribution unit mainly constitutes main, sub – main and lateral lines.

Main line

The main line conveys water from the filtration unit to sub main. Sub main conveys water from main line to the laterals. Main and sub mains are usually PVC pipes. The PVC pipes are available in the market with external diameter sizes of 20 mm, 25 mm, 30 mm etc. Usually 2 kg/cm^2 to 4kg/cm^2 strength of pipes are used for irrigation. 2 kg /cm^2 pipes can withstand 20 m height water column and 4 kg/cm^2 pipe can withstand 40 m height of water column. The PVC pipes have to be buried under the land. It should not be laid over the ground. If it is laid over the ground, it will become brittle and get broken in a few years. Sometimes High Density Poly Ethylene (HDPE) pipes are also used. They can be laid over the land. But they are costly compared to the PVC pipes.

Sub main

It conveys water from main line to the laterals. Rigid PVC pipes are commonly used as sub main and recently better quality HDPE is being used. Size of the pipe depends on water requirement of the cop and size of the field. The outer diameter ranges from 2.5 to 4 cm with pressure rating of 4 kg/cm^2. There may be many number of sub mains depends on the plot size and crop. These pipes are buried under 60-90 cm or below ground.

2.2.3 Water delivery unit

Laterals

Laterals are always made of Low density Poly Ethylene (LDPE) pipes. Laterals are small diameter flexible pipes or tubing which are laid along the rows of the crop on which emitters are connected directly or through micro tube to provide water to the emitters. The lateral pipe size of 10mm to 20 mm with wall thickness varying from 1 to 3 mm and a pressure rating of 4 kg/ cm^2. The laterals are usually laid over the ground. Holes are drilled in the sub mains with drill bits. Then a start washer is inserted in to the hole and a start connector is driven into the washer using a wooden hammer. The lateral pipe is inserted over the start connector. The laterals are closed with end line by folding the lateral pipes at the ends.

Emitter / Drippers

Emitters are the devices, through which water comes out of the pipe network. The emitter is metering device made of plastic that delivers a small but precise discharge. The quantity of water released from this emitter is usually expressed in liter per hour.

Types of drippers

The drippers may be classified into two based on the way by which the drippers are attached to the later pipes.

- Online drippers
- Inline drippers

Online drippers

Online drippers are attached on the lateral by drilling a hole and inserting into the lateral by hand. Or, a micro tube start will be inserted into the hole and micro tube will be attached to the micro tube start and dripper will be attached at the end of the micro tube. This kind of micro tube connections are needed when many drippers are provided to each tree. All online drippers can be opened up for cleaning.

Inline drippers

Inline drippers are inside the lateral pipes. They are integral part of laterals. The spacing between them is fixed while manufacturing itself; with different spacing and dischrates they are available in the market. They are very much useful in row field crops and small tree crops like citrus, drumstick etc.

Accessories and fittings

1. Control valve
2. Flush Valve
3. Air release vacuum breaker
4. Non - Return Valve
5. Pressure Gauge
6. Regulators
7. Main and Sub main fittings
8. Tee Connector
9. Straight connector or joiner
10. Gromet and Take off
11. End Cap

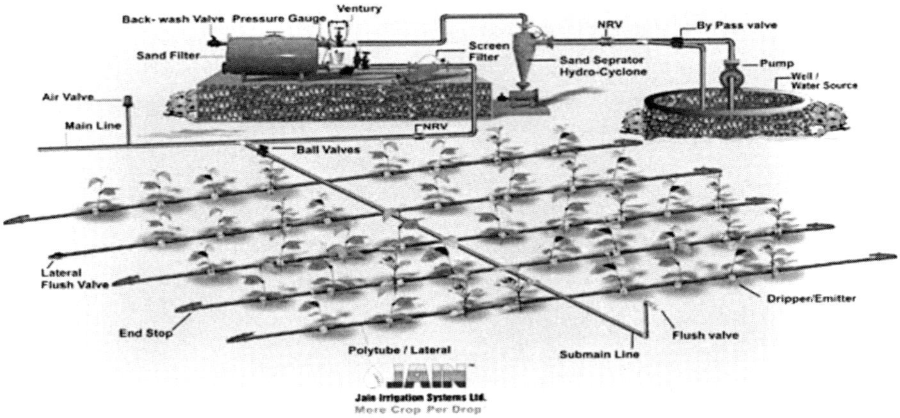

Fig. 12. A typical layout of drip irrigation system

3. SPRINKLER IRRIGATION

Sprinkler Irrigation refers to application of water to crops in the form of spray from above the crop like rain. It is also called the overhead irrigation as water is allowed to fall as spray from above the crop. Sprinkler irrigation also called as pressurized irrigation as water under pressure is carried and sprayed into the air above the crop through a system of pipe network.

The sprinkler (overhead or pressure) irrigation system conveys water to the field through pipes (aluminium or PVC) under pressure with a system of nozzles. This system is designed to distribute the required depth of water uniformly which is not possible in surface irrigation. Water is applied at a rate less than the infiltration rate of the soil hence the runoff from irrigation is avoided.

3.1 Sprinkler system classification

Based on the portability

(i) **Portable system:** It has portable mainlines and laterals and a portable pumping unit

(ii) **Semi portable system:** A semi portable system is similar to a fully portable system except that the location of the water source and pumping plant are fixed.

(iii) **Semi permanent system:** A semi permanent system has portable lateral lines, permanent main lines and sub mains and a stationery water source and pumping plant. The main lines and sub mains are usually buried, with risers for nozzles located at suitable intervals.

(iv) **Solid set system:** A solid set system has enough laterals to eliminate their movement. The laterals are placed in the field early in the crop season and remain for the season.

(v) **Permanent system:** It consists of permanently laid mains, sub mains and laterals and a stationary water source and pumping plant. Mains, sub mains and laterals are usually buried below plough depth. Sprinklers are permanently located on each riser.

Advantages

- Water saving to an extent of 35-40% compared to surface irrigation methods.
- Saving in fertilizers - even distribution and avoids wastage.
- Suitable for undulating topography (sloppy lands)
- Reduces erosion
- Suitable for coarse textured soils (sandy soils)
- Frost control - protect crops against frost and high temperature
- Drainage problems eliminated
- Saving in land
- Fertilisers and other chemicals can be applied through irrigation water

Adaptability of sprinkler system

Sprinkler irrigation may be used for many crops and on all types of soil on lands of varying topography and slopes. However it is best suited to irrigate.

1. Sandy soils and soils having high infiltration rate
2. Shallow soils that do not allow proper land leveling required for surface irrigation methods.

3. Areas with steep slopes having erosion hazards.
4. For growing high priced crops
5. Where water is scarce and costly.

Suitable slopes

Sprinkler irrigation is adaptable to any farmable slope, whether uniform or undulating. The lateral pipes supplying water to the sprinklers should always be laid out along the land contour whenever possible. This will minimize the pressure changes at the sprinklers and provide a uniform irrigation.

Suitable soils

Sprinklers are best suited to sandy soils with high infiltration rates although they are adaptable to most soils. The average application rate from the sprinklers (in mm/hour) is always chosen to be less than the basic infiltration rate of the soil .so that surface ponding and runoff can be avoided. Sprinklers are not suitable for clay soils which easily form a crust. If sprinkler irrigation is the only method available, then light fine sprays should be used. The larger sprinklers producing larger water droplets are to be avoided.

Suitable irrigation water

A good clean supply of water, free of suspended sediments, is required to avoid problems of sprinkler nozzle blockage and spoiling the crop by coating it with sediment.

Selecting the most appropriate sprinkler systems

While selecting a sprinkler system, the most important physical parameters to be considered are:

1. The crop or crops to be cultivated.
2. The shape and size (acres) of the field.
3. The topography of the field.
4. The amount of time and labour required to operate the system.

3.2 A sprinkler system usually consists of the following parts

1. Pumping unit
2. Debris removal equipment
3. Pressure gauge / water-meter
4. Pipelines (mains - sub-mains and laterals)
5. Couplers
6. Raiser pipes
7. Sprinklers
8. Other accessories such as valves, bends, plugs, etc.

Pumping unit

A high speed centrifugal or turbine pump can be installed for operating the system for individual farm holdings. The pumping plants usually consist of a centrifugal or a turbine type pump, a driving unit, a suction line and a foot valve.

Pipe lines

Pipelines are generally of two types. They are main and lateral. Main pipelines carry water from the pumping plant to many parts of the field. In some cases sub main lines are provided to take water from the mains to laterals. The lateral pipelines carry the water from the main or sub main pipe to the sprinklers. The pipelines may be either permanent, semi permanent or portable.

Couplers

A coupler provides connection between two tubing and between tubing and fittings.

Sprinklers

The sprinkler heads may have single nozzle or twin nozzles. Single nozzle sprinklers wet smaller diameters (3.5 m – 14.0 m). Twin nozzle sprinklers wet larger diameters. In these sprinklers, one nozzle is smaller and another nozzle is larger. Smaller nozzle would wet areas nearer to the centre and smaller nozzle would wet peripheral areas. Rain guns are capable of wetting very large diameters as much as 100 m. Sprinklers are designed to rotate without any external agency. Water flowing through larger nozzle, hits on a rocker arm and this rocker arm provides an impact and rotates the sprinklers. These types of sprinklers are called as impact sprinklers. Rain guns and large sprinklers will have a gear inside the sprinkler head for rotating the sprinkler. They are called as gear type and the rotation of the gear type is very smooth compared to impact type.

Other accessories / fittings

- **Water meters:** It is used to measure the volume of water delivered.
- **Pressure gauge:** It is necessary to know whether the sprinkler is working with the desired pressure in order to deliver the water uniformly.
- Bends, tees, reducers, elbows, hydrants, butterfly valves, end plugs and risers
- **Debris removal equipment:** This is needed when water is obtained from streams, ponds, canals or other surface supplies. It helps to keep the sprinkler system clear of sand, weed seeds, leaves, sticks, moss and other trash that may otherwise plug the sprinklers.
- **Fertilizer applicators:** These are available in various sizes. They inject fertilizers in liquid form to the sprinkler system at a desired rate.

4. COMPARISON OF DRIP, SPRINKLER AND SURFACE IRRIGATION

Drip	Sprinkler	Surface irrigation
Less water required	Comparatively more water required	More water required
Saline water can be used	Saline water sprinkling is harmful	Saline water cannot be used
Most suitable for widely spaced tree crops and high value crops	Saline for close spaced crops	Suitable for closely spaced crops
Low pressure system required	High pressure system required	Medium pressure pump set required
No evaporation loss	20 - 30% evaporation loss	15 - 20% evaporation loss
Only required area near the crop is irrigated	Unwanted areas are also sprinkled	Entire surface area is wetted
Less weed growth	Comparatively more lob our required	More lab our required
Can be operated during night hours	Cannot be operated during night time	Cannot be operated at night hours
Less weed growth	More weed growth as entire area irrigated	More weed growth
Plant protection chemicals can be applied during from irrigation	Chemicals cannot be applied during sprinkling	Chemicals cannot be applied during irrigation
Drip system almost free from corrosion	In saline water, pipes are corroded	Corrosion does not arise
Applied water remain constant in root zone	Periodical sprinkling water never constant	Soil moisture is fluctuated
Soil moisture is maintained at field capacity	Economically impractical to maintain soil moisture at field capacity	Practically impossible to maintain uniform moisture lever at root zone
Annual electricity charges are less	Electricity charges are high	Electricity charges lower than sprinkler
Wind has no effect	Wind affects uniformity of water application	No wind effect

5. RAIN GUN IRRIGATION

Rain gun irrigation is a type of sprinkler irrigation and covers larger area in a short time and suitable for all the crops particularity for high statured crops like sugarcane, tapioca, cotton and orchards. Most useful in situations where the supply of labour is scanty or expensive. Large quantities of water are sprayed on an increasing scale with the help of towers spread out inside the field.

Rain gun sprinklers operate under low, medium and high pressure systems, rotating at uniform speed and can be adjusted to cover part circle or full circle. The entire sprinkler is driven by gearboxes housed inside the body made up of pressure die cast aluminum. The rain gun is perfectly balanced and has two discharge outlets. Interchangeable nozzles of different sizes can be fixed depending on the dead length and static head.

Advantages

- Highly energy efficient, Long lasting and trouble free
- Reduces labour cost
- The precipitation varies with the nozzle size, but unlike conventional sprinklers, overlapping is not occur and this reduces the time required for irrigation as well as the savings on fuel.
- Ease installation, assembly and maintenance
- Fine droplet adjustment, like mist is helpful in creating a microclimate.
- Rain gun irrigation results in improved efficiency compared to conventional sprinklers.

Disadvantages

- High day time temperatures result in evaporation loss.
- Not suitable in areas with high velocity Winds.
- Prohibitive cost of rain guns.

6. MANAGEMENT AND MAINTENANCE OF MICRO IRRIGATION SYSTEM

For proper operation and maintenance of drip irrigation systems, the following parameters have to be tested. Many of these parameters are useful in the selection of filters. Certain parameters are useful in the operation and maintenance.

Criteria for plugging potential of micro irrigation water sources			
	Plugging Hazard Based on Concentration		
	Concentrations		
Factor	**Slight**	**Moderate**	**Severe**
Physical			
Suspended solids (filterable) ppm	< 50	50 to 100	> 100
Chemical			
pH	< 7.0	7.0 to 7.5	> 7.5
Dissolved solids (ppm)	< 500	500 to 2000	> 2000
Manganese (ppm)	< 0.1	0.1 to 1.5	> 1.5
Iron (ppm)	< 0.1	0.1 to 1.5	> 1.5

Hydrogen sulfide (ppm)	< 0.5	0.5 to 2.0	> 2.0
Hardness* (ppm)	< 150	150 to 300	> 300
Biological			
Bacteria (population)	< 10,000	10,000 to 50,000	> 50,000
*Hardness as ppm $CaCO_3$			

All the parameters need not be studied for all situations. If the source of water is surface water like rivers and lakes, mostly biological impurities like algae and bacterial slime are predominant. In such cases, other tests need not be carried out.

When water is taken from deep wells, the presence of Hydrogen Sulfide is very less. In such cases, this test can be avoided. The presence of hydrogen sulfide can be easily known if the water has rotten egg smell.The presence of Iron can be easily known if the water has a brown colour.

6.1 Mechanism behind plugging of emitters

6.1.1 Chemical reactions

Water is often referred to as the universal solvent since almost everything is soluble in it to some extent. The solubility of any material in water is controlled by variations in temperature, pressure, pH, redox potential and the relative concentrations of other substances in solution. Three gases (oxygen, carbon dioxide, and hydrogen sulfide) are important in determining the solubility characteristics of water. These gases are very reactive in water, and they determine to a significant extent the solubility of minerals within a given water source.

Carbon dioxide gas is of particular importance in the dissolution and deposition of minerals. Water absorbs some CO_2 from the air, but larger quantities are absorbed from decaying organic matter as water passes through the soil. Under pressure, as is in groundwater, the concentration of CO_2 increases to form carbonic acid. This weak acid can readily dissolve mineral compounds such as calcium carbonate to form calcium bicarbonate which is soluble in water. This process allows calcium carbonate to be dissolved, transported in the irrigation water. The bicarbonates under certain conditions again get deposited on pipes and emitters as calcium carbonate.

In the wells with Air Lift pumps, since the water is lifted by supplying air to the well water, the amount of CO_2 dissolved is very high. Therefore, the amount of calcium bicarbonate dissolved from the subsurface formations is also high. Therefore, in air lift pumps, the pipes get deposition of calcium carbonates significantly

Chemical plugging usually results from precipitation of one or more of the following minerals: calcium, magnesium, iron, or manganese. The minerals precipitate from solution and form encrustations that may partially or completely block the flow of water through the emitter. Water containing significant amounts of these minerals and having a pH greater than 7 has the potential to plug emitters. Particularly common is the precipitation of calcium carbonates, which is temperature and pH dependent. An increase in either pH or temperature reduces the solubility of calcium in water, and results in precipitation of the mineral.

When groundwater is pumped to the surface and discharged through a micro irrigation system, the temperature, pressure, and pH of the water often changes. This can result in the precipitation of calcium carbonates or other minerals to form scale on the inside surfaces of the irrigation system components.

Iron is another potential source of mineral deposit that can plug emitters. Iron is encountered in practically all soils in the form of oxides, and it is often dissolved in groundwater as ferrous bicarbonate. When exposed to air, soluble ferrous bicarbonate oxidizes to the insoluble or colloidal ferric hydroids and precipitates. The result is commonly referred to as 'red water,' which is sometimes encountered in wells. Manganese will sometimes accompany iron, but usually in lower concentrations.

Presence of Hydrogen sulfide will minimize the precipitation of calcium carbonate ($CaC0_3$) because of its acidity. But if Iron is present, Iron sulphate is precipitated.

6.1.2 Bio-chemical reactions

Algal and bacterial growth is major problems associated with the use of surface water. Whole algae cells and organic residues of algae are often small enough to pass through the filters. These algal cells form aggregates that plug emitters. At the exit of the dripper, due to sunlight, the plugging is very fast. Residues of decomposing algae can accumulate in pipes and emitters to support the growth of slime-forming bacteria. Chemical precipitation is normally not a major problem when using surface water.

Certain bacteria can cause enough precipitation of manganese, sulfur, and iron compounds to cause emitter plugging. In case of Iron, soluble ferrous in the water is oxidized by certain bacteria into insoluble ferric form. Hydrogen sulfide is oxidized by some bacteria into elemental Sulphur slimes.

6.1.3 Chemical reactions of fertilizers with irrigation water

In water with calcium, magnesium and bicarbonates, if phosphate containing fertilizers are mixed, calcium and magnesium phosphates are formed. By reducing pH, this effect can be reduced.

If polyphosphate fertilizers are mixed with water with Calcium and Magnesium ions, gel suspensions are formed which clog filters and drippers.

When water contains Calcium and if sulphate fertilizers are mixed with it, Calcium sulphate is precipitated. With rise in temperature, this problem would get aggravated.

When water contains Calcium and bicarbonate ions, mixing with Urea precipitates Calcium Oxide commonly known as lime.

When Phosphoric acid is mixed with water containing Calcium, Calcium Phosphate is formed.

If water contains more than 0.1 ppm of Iron, and if Calcium or Phosphate fertilizers are injected, then iron precipitate is formed.

6.2 Management of plugging/clogging

6.2.1 Chlorination

If the source of irrigation water is dam, river, irrigation channel, water would contain organic matter (algae, bacterial slime). In such situations, chlorination is inevitable. Chlorine is an oxidizing agent which kills bacteria, algae and organic matter. Normally used chlorine compounds are sodium hypochlorite (liquid) and calcium hypochlorite (solid). Sodium hypochlorite contains 10% free available chlorine. Calcium hypochlorite ($Ca(OCl)_2$) contains 70 % free available chlorine. Bleaching powder contains only 35 % of free available chlorine. Venturi or fertilizer tank can be used for injecting the dissolved chlorine compounds into the water.

There are 3 methods of chlorine application.

1. Continuous treatment
2. Superchlorintion
3. Intermittent treatment

Continuous treatment

It is the best and simplest method. In long run, it provides good results as this method continually removes the organic growth and the system is all the time clean from organic material. The required concentration varies according to the water quality. Normal injection rate is 1 ppm and to begin with 5-10 ppm can be adopted.

Intermittent treatment

When less organic matter exists in the water, 10-20 ppm injection for 30 minutes at the end of irrigation can be done. Chlorine is left in pipes till next irrigation. This is called as intermittent method.

Super chlorination

Without chlorination if the irrigation system is operated for some time, there is possibility of high accumulation of organic matter. This may completely clog the system. In such situations, 500 ppm is injected and the system is shut down and left for up to 24 hours. Then the system is flushed step by step. First mainline, then submains and laterals. Do not flush directly through the drip laterals. This will clog the emitters in the drip line.Super-chlorination may damage sensitive plants and irrigation system components.

Exactly finding the correct dosage of chlorine is difficult because the chlorine demand of each and every water is different. Use a trial and Error method to achieve a specific free residual chlorine concentration. At least 0.5 ppm of free chlorine at the end of laterals ensures that all the organic matter has been destructed. Use a field test kit to measure chlorine concentration from water obtained from drippers at the farthest end.

Hydrogen sulfide consumes about 2 ppm chlorine for each ppm of sulfide, while iron consumes about 0.7 ppm chlorine for each ppm of iron.

Example

- Pump discharge – 2 lps
- Concentration required – 5 ppm
- Irrigation time – 3 hours
- Calculate Sodium Hypochlorite – 10 % Chlorine

  Total Volume of water = 2 lps × 3 × 3600 = 21600 litres

  Chlorine required = 5 × 21600 = 108000 mg = 108 gm

  Sodium hypochlorite required = 108 × 10 = 1080 gm

6.2.2 Acid treatment

Addition of acid to irrigation water is done for the following situations:

- Lowering the water pH can enhance the effectiveness of chlorine.
- The pH-lowering power of acid can prevent precipitation of solid compounds, particularly calcium carbonate ($CaCO_3$).
- Citric acid prevents iron scale formation when continuously injected at 25 ppm.

It is always recommended to neutralize 80% of the bases (carbonates and bicarbonates) in the water to eliminate carbonate precipitation.

Acids normally used are as follows:

- Hydrochloric acid (32% Strength)
- Sulfuric acid (93 % strength)
- Phosphoric acid (85 % strength)

Sulphuric acid is highly toxic and hence it is not used popularly. Phosphoric acid should not be used if there is more than 50 ppm Ca in the irrigation water because calcium phosphate will precipitate. Calcium phosphate is nearly insoluble and does not readily dissolve. Hydrochloric acid is very much commonly used.

Acid treatment for prevention of carbonate scales

Acid treatment can be continuously done to prevent the scale formation due to carbonate salts in the water. If these salts plug the emitters and pipe, reclamation by acid treatment becomes difficult. It is because the emitters will be discharging the water and it takes more time for removal of the salts. Hence, if the carbonate content is more than a critical level continuous treatment with acid for all irrigations is recommended.

The best way to find out the acid dose for any irrigation water is to titrate a specific volume of irrigation water to bring its pH to 7. For safety, always use pH paper and test the pH of irrigation water after acidification. If the pH is different from the desired pH, then titration should be once again done as the quality of irrigation water may change dynamically.

Acid treatment for clogged systems

When systems get either partially clogged or fully clogged, it becomes difficult to reclaim the system. The pH of the water should be brought down to the range of 4 to 2. It should be borne in mind that as the pH is reduced more and more, corrosion of various components of irrigation system would also occur. So, the pH reduction should be planned according to the requirement. Generally it is recommended that 1 litre of HCl should be done for 1000 litres of water. The acid treatment should be done for half an hour. The acid treatment should be done at the end of any irrigation. The water in the pipes should be left as such for overnight and next day the system should be flushed. During flushing first main lines should be flushed and then laterals. Otherwise, there will be possibility of emitters getting clogged with the removed scales.

Let us assume that the drippers are discharging at an average rate of 1.5 lph and 2000 drippers are simultaneously operating. Then the total discharge rate of the system is 3000 lph. If the acid treatment is done for half an hour, the total discharge from the system would be 1500 litres. Therefore, as per the preceding general recommendation, the amount of HCl needed is 1.5 litres.

Let the venturi suck at the rate 50 lph. So, for an half an hour, it can suck 25 litres. Therefore, dilute 1.5 litres of acid into around 25 litres and place under the venturi.

6.2.3 Algae control in ponds and wells

Copper sulphate can be used for controlling the algae growth in the ponds and wells. But if the water is used for domestic purposes and also for animals, it is better not to use it. Copper sulphate can be taken in porous cloth or gunny bag and tied in such a way that the salt is floating on the surface of the water. The concentration of copper sulphate may be in the range of 0.5 to 2 ppm. Following table provides the safe levels of concentration that does not affect fish life.

Copper sulfate ($CuSO_4$) levels safe for fish	
Alkalinity Value ($CaCO_3$, mg/l)	**Addition of Copper Sulfate**
below 40	do not use
40-60	0.4 gm/ m^3 of water
60-100	0.5 gm/ m^3 of water
over 100	1 gm/ m^3 of water

6.2.4 Remedy for iron

If the primary cation is iron, removal and cleaning or replacement with new emitters is likely the only solution. Iron compounds found in plugged emitters are very difficult to unplug in the field. Plugged emitters can be cleaned by soaking with Citric acid (0.5-1.0%) solution for 24 to 48 hours. For controlling the iron scales, 1 mg/l of chlorine or 1 % citric acid can be injected with all the irrigation.

A very simple method of controlling iron problem is to pump the water to a surface tank and aerate sufficiently so that the dissolved bicarbonate gets precipitated into ferric forms.

6.2.5 Remedy for manganese

Similar to Iron, for manganese also, if water is aerated, manganese gets precipitated. Polyphosphate and Polymaleic acid injection has been found to be good for removing manganese scales.

6.2.6 Remedy for hydrogen sulphide

The processes mentioned in preceding sections namely, Aeration, Chlorination and Acid treatment have been found to be useful for the problem of Hydrogen Sulphide remedy.

6.3 Regular maintenance of accessories

For the successful maintenance of drip irrigation systems, maintenance of various devices are very important.

6.3.1 Filters

Examine screens and seams of filters for damages. Proper tightness between screen, cover and body of the filter prevent water leaks. For disc type filters, disc assembly should be properly tightened. The maximum Allowable pressure differential between inlet and outlet of the filter is 5 meters. When the pressure differential is above 5 m flushing and cleaning should be done. It is always better to flush the filters before each irrigation. Avoid entry of dirt into pipe, when opening a filter.

6.3.2 Gravel filters

Rinse gravel filters from back with open top cover under controlled pressure to avoid gravel escape. For removal of the aggregates and algae, fill container with chlorine and allow soaking for 24 hours. After rinsing, pour off the water and leave the filter in a dry and ventilated place.

6.3.3 Flushing main &laterals

During installation process, before connecting laterals flush main and sub main. During failure or burst of the main lines, flush dirt from lines for at least 10 minutes.

Filtration does not prevent very small particles and water soluble matter from going into the pipe network. A layer of sediment is deposited on the inner walls of the pipe and accumulation occurs more at the end of laterals. Therefore flush the laterals after 2 to 4 weeks.

6.4 Handling rats and squirrel problems

Rats and squirrel bite the laterals. This problem is very difficult to solve. One way is directly controlling rats by chemicals. The drip lateral production companies claim that they produce laterals with a rat repellants. The effectiveness of these materials can be reported only after they are field tested sufficiently.

7. GOVERNMENT SUPPORT TO POPULARIZE MICRO IRRIGATION SYSTEMS

The Government of India has been implementing Centrally Sponsored Scheme on Micro Irrigation with the objective to enhance water use efficiency in the agriculture sector by promoting appropriate technological interventions like drip & sprinkler irrigation technologies and encourage the farmers to use water saving and conservation technologies.The Scheme was launched by the Department of Agriculture & Cooperation, Ministry of Agriculture in January, 2006 as Centrally Sponsored Scheme on Micro Irrigation (CSS). In June, 2010, it was up-scaled to National Mission on Micro Irrigation (NMMI), which continued till the year 2013-14.

From 1st April, 2014, NMMI was subsumed under National Mission on Sustainable Agriculture (NMSA) and implemented as "On Farm Water Management" (OFWM) during the financial year 2014-15. From 1st April 2015, Micro Irrigation component of OFWM has been subsumed under Pradhan Mantri Krishi Sinchayee Yojana. It will be implemented as Centrally Sponsored Scheme on Micro Irrigation during the financial year 2015-16 as per the same pattern of assistance and cost norms as were prevailing under OFWM, until revised.

8. PMKSY (PER DROP MORE CROP - PDMC)

Department of Agriculture, Cooperation & Farmers Welfare is implementing Per Drop More Crop component of PMKSY, which is operational from 2015-16 in the country. Per Drop More Crop mainly focuses on water use efficiency at farm level through precision/ Micro Irrigation (MI) (Drip and Sprinkler Irrigation). Besides promoting precision irrigation and better on-farm water management practices to optimize the use of available water resources, this component also supports gap filling interventions as Other Interventions like micro level water storage or water conservation/management activities to complement and supplement the source creation. Budget for the Year 2019-20, budget provision of Rs. 3500 crores has been made to implement the Per Drop More Crop component.

9. MICRO IRRIGATION FUND

A dedicated Micro Irrigation Fund created with NABARD has been approved with an initial corpus of Rs. 5000 crore (Rs. 2000 crore for 2018-19 & Rs. 3000 crore for 2019-20) for encouraging public and private investments in Micro irrigation. The main objective of the fund is to facilitate the States in mobilizing the resources for expanding coverage of Micro Irrigation. MIF would not only facilitate States in incentivizing and mobilizing resources for achieving the target envisaged under PMKSY-PDMC but also in bringing additional coverage through special and innovative initiatives by State Governments.

QUESTIONS

I. Fill in the blanks

1. Low water emission rates and localized delivery of water to a limited area is known as __________ irrigation
2. Drip irrigation concept developed by ____________
3. Drip irrigation suitable for __________ land
4. ___________ valve arrest the reverse motion water in the micro irrigation systems
5. ___________filters are efficient in filtration of organic material and algae in micro irrigation

II. Choose the best answers

1. Precise, slow and continuous application of water in the form of discrete drops through emitter is called as _____________ irrigation
   (a) Sprinkler (b) Drip
   (c) Micro jet (d) None of the above
2. Discharge rate of emitters in line source and point source are less than ________ & ______ respectively
   (a) 4 l/hr & 8 l/hr (b) 3 l/hr & 9 l/hr (c) 8 l/hr & 4 l/hr (d) 2 l/hr & 8 l/hr
3. Waterways of emitters ranging from __________ mm in diameter
   (a) 0.5 – 1.0 mm (b) 0.4 – 2.2 mm
   (c) 0.2 – 2.0 mm (d) None of the above
4. ___________ helps to inject the fertilizer solution into the irrigation water in Drip system
   (a) Non return valve (b) Venturi
   (c) Pressure gauge (d) A & B
5. __________ refers to application of water to crops in the form of spray from above the crop like rain
   (a) Drip irrigation (b) Sprinkler irrigation
   (c) Jet irrigation (d) A & B

III. Answer the following questions

1. Define: Micro irrigation, Drip Irrigation, Sprinkler irrigation
2. Draw the neat sketch of drip and sprinkler irrigation
3. Give the components of drip and sprinkler irrigation
4. Give details on micro irrigation maintenance.
5. List out the advantages and disadvantages of various micro irrigations.
6. Types of drip irrigation

7. Write the drip irrigation components and function
8. Types of filter used in micro irrigation system and advantages of filters
9. Explain water distribution and delivery unit of drip irrigation
10. Write the types of drippers and uses
11. Classification of sprinkler system
12. Differentiatedrip, sprinkler and surface irrigation methods
13. What is rain gun irrigation? and write advantages and disadvantages
14. PMKSY
15. Mechanism behind plugging of emitters
16. Management of plugging/clogging of delivery unit

6

WATER MANAGEMENT FOR CEREALS AND PULSES

1. TRANSPLANTED RICE

Rice is a semi-aquatic plant and its water requirements are much higher than most crops. About 40 to 70% of water diverted for irrigation of rice is lost by deep percolation. Water is very inefficiently used in rice production.

Submergence Vs Non-submergence

Submergence means standing water in rice field or land submergence is common practice in lowland rice. Rice has ability to grow in land submerged conditions is due to unique mechanism it has for transmitting oxygen from the air through leaves to its roots for respiration. Rice grown under submergence gives even 50% more yield than grown without submergence.

Shallow depth of submergence of soil surface (about 2.5 cm water) throughout the crop growth period is conducive to higher yields. Continuous submergence has some advantages: less weed problem, fixation of nitrogen due to blue-green algae, increased availability of plant nutrients especially P, Fe, K, Ca and silicon, regulation of soil temperature, better crop growth, higher yields and reduction in labour cost.

Rice can be grown if soils are saturated with water or thin film of water is retained on the soil. But, lot of care is necessary to keep the soil always at this level. Shallow depth of water leads to considerable saving in water and increases rice yield.

Depth of submergence

It is not entirely necessary to submerge the land continuously to a depth of 5 cm. After basal application of fertilizers and incorporation, the soil has to be flooded with shallow layer of 2 cm depth of water. The fertilizer has to be placed below 5 cm depth to get high grain yields. At transplanting also, 2 cm depth of water will help in shallow planting of seedling. 5 cm depth of water is necessary for 3 days after transplanting for quick establishment of seedlings.

About 5 cm depth of water is maintained from the beginning of panicle initiation to 21 days after flowering and irrigation is withheld subsequently. The crop does not require any irrigation from 15 days before harvest maturity. The moisture sensitive periods for rice are panicle initiation, heading and flowering. During these periods, it is essential to have land submergence up to 5 cm.

Standing Vs flowing irrigation

Irrigation efficiency is higher when water stands (stagnates) in the rice field than in flowing irrigation. But, flowing irrigation is necessary when harmful salts are present in soil or water, to reduce water and soil temperature and where labour is expensive and scarce. Water use efficiency is decreased by about 30% and yield by about 8% if flowing irrigation is practiced throughout the crop period.

Intermittent irrigation

Under high humidity and low evaporative demand, the practice of intermittent submergence i.e. submergence during the critical stages of crop (i.e. initial tillering, panicle initiation and flowering) and maintenance of saturation or drying up to hair cracking stage during the rest of the crop stages has given grain yields comparable to continuous shallow submergence.

Water requirement

The average water requirements are: Nursery - 40 mm; main field preparation - 200 mm and irrigation for the crop - 1000 mm.

2. DIRECT SEEDED RICE

Direct seeded rice on puddled soil required about 30% more water than transplanted rice though the period of maturity of the crop was same under both conditions.

3. TECHNIQUES OF IRRIGATION IN RICE

Both surface and pressurized methods of irrigation are practiced in rice cultivation.

Surface irrigation

Flooding; and beds and channels methods are commonly practiced in surface method of irrigation of rice.

Flooding

Under this method, field is puddled well and either seeds are sown or seedlings are transplanted. The water is allowed to stagnate most of the life cycle of rice. This is common in irrigated lowland condition. Here, saturation of soil is the pre-requisite for better cultivation of rice.

Beds and channels

Under rainfed upland and low land situations, the rice is cultivated like other upland crops and water is allowed through surface channels. The frequency of irrigation depends on soil and climatic factors. The yield levels are comparatively low in this type of cultivation.

Ridges and furrows are also formed instead of beds and channels in few a cases.

Pressurized irrigation

Rice is raised under rainfed condition and water is allowed through drip system. Here, about 125% of ET is considered for better growth and yield. About 60-75% water is consumed compared to the flooding system and yield levels also low.

4. WHEAT

The dwarf wheat varieties were first introduced into India during 1963. In the very first year of their cultivation, it was realized that the irrigation practices developed for the tall wheat varieties did not suit the dwarf wheat varieties. Consequently, studies to determine the optimum irrigation & water requirements of dwarf wheat were taken up.

Optimum time of first irrigation

The crown root initiation (CRI) stage (about 3 weeks after sowing) is the optimum time for first irrigation in dwarf wheat. A delay of 2 – 3 weeks in this irrigation reduces the yield by 500 – 1000kg/ha.

Optimum number of irrigations

If the underground water table is well below the root zone and there is no rainfall during the growing season, 4-6 irrigations depending on the climatic conditions, soil type and the variety may be needed for getting the highest yield from the dwarf wheat.

Based on soil water regime approach under south Indian condition, irrigation at 25% ASM and 40-50% ASM under north Indian conditions were found to be optimum. Likewise, irrigations at 0.75 to 1.0 IW/CPE ratio based on climatological approach were found to be optimum for many locations. An IW/CPE ratio of 1.0 was optimum at CRI stage. The critical growth stages for irrigation are crown root initiation, tillering, heading & grain formation. The seasonal crop water requirement varies between 350 to 500mm depending upon the agro-climatic zone. The recommended method of irrigation is check basin and border strip method of irrigation and sometimes corrugation method of irrigation.

5. MAIZE

For maximum production a medium duration grain crop of maize requires between 400 to 600 mm of water depending upon the soil-plant and climatic factors. The furrow method of irrigation is commonly practiced in India.

Frequency and depth of irrigation has a pronounced effect on grain yield. Maize appears relatively tolerant to water deficits during the vegetative and ripening periods. Greatest decrease in grain yield is caused by water deficits during the flowering period including tasseling, silking and pollination, mainly due to a reduction in grain number per cob. Water deficits particularly at the time of silking & pollination may result in little or no grain yield due to silk drying. Water deficits during the ripening period have marginal effect on grain yield. To obtain good stand and rapid root development, the root zone should be wetted at or soon after sowing.

Taking into account the level of evaporative demand to meet full water requirement irrigations have to be scheduled at 40% ASM in the germination and seedling & establishment stages. Whereas, during vegetative, tasseling, silking, pollen shed and kernel development stages irrigations have to be scheduled at 55 to 65% ASM and up to 80% ASM during ripening or maturation period. Likewise, irrigation at an IW/CPE ratio of 1.0 at critical growth stages and 0.5-0.75 at other stages was found to be optimum.

The threshold soil moisture tension levels for scheduling irrigations based on tensiometer readings vary between 0.3-0.6 bars at different growth stages. Under conditions of marginal rainfall and limited irrigation water supply, the number of possible irrigation applications may vary between 2–5.

6. SORGHUM

Sorghum is a rainfed crop grown both in red and black soils. It is cultivated in rainy, winter and summer seasons. It is a drought tolerant crop respond well to irrigation. Water requirement varies from 450-600mm.Irrigations at 50% ASM or IW/CPE ratio of 0.6 at the above four stages is adequate for normal yield.

Seedling, panicle initiation, booting (panicle inside the leaf sheath of flag leaf), flowering and grain formation stages. Of these, early seedling and flowering are most critical.

In shallow soil, it is important to maintain surface moisture seedling stage for 15 days after planting to encourage the development of strong deep root system. An irrigation 15 days after emergence is necessary. Irrigation prior to heading and 10 days after heading is necessary.

7. PEARL MILLET

Pearl millet is mainly grown as a rainy season crop under rainfed conditions. In small pockets it is grown as summer crop under irrigated conditions. For scheduling irrigation, 50% of available soil moisture is required.

It is drought resistant and also sensitive to water logged conditions. Grow well in soils having high salinity or low pH.Water requirement is 250 to 300 mm for high yield. Irrigation requires during moisture sensitive period.

This crop is irrigated by furrow method. This crop is grown as transplanted crop in summer. Heavy irrigation is given and 20 days old seedlings are transplanted. A light irrigation is given after 4 days of transplanting for better establishment. Second irrigation after 10-15 days. Third irrigation after hand hoeing. Subsequently, irrigation is scheduled at weekly intervals. Irr1igation can be with held 10 days before harvesting.

8. FINGER MILLET

Finger millet is grown in South India both in rainy and summer seasons. Summer season crop is grown under irrigation. Irrigation is scheduled at 50% of depletion of available soil moisture. IW/CPE ratio is 0.6-0.8. This crop is sensitive to water logging.

Water requirement is 300-400mm. This crop is usually cultivated under check basins. Panicle initiation and flowering are critical period for irrigation.

Pre-planting irrigation is given for transplanting seedlings. Irrigation is usually given on third or fourth day after planting for better establishment. Water is then hold for 10-15 days for hardening of seedlings.

9. MINOR MILLETS

9.1 Pani varagu (Proso millet)

Proso millet sown during Kharif season, generally does not require any irrigation. However, at tillering stage, if dry spell prevails for longer period, then one irrigation must be given to boost yields. Summer crop, however, would require two to four irrigations depending upon soil type and climatic conditions. Give first irrigation 25-30 days after sowing and second irrigation about 40-45 days after sowing. Due to shallow root system of proso millet, heavy irrigation is not advisable.

9.2 Kudiraivali (barnyard millet)

Generally barnyard millet does not require any irrigation. However, if dry spell prevails for a long period, then one irrigation must be given at the time of panicle initiation stage. It is always better if excess water of heavy rains is drained out of the field.

10. PULSES

10.1 Redgram

Irrigate immediately after sowing, 3rd day after sowing, bud initiation, 50% flowering and pod development stages. Water stagnation should be avoided. Water requirement is 300mm.

10.2 Soybean

Irrigate immediately after sowing. Give life irrigation on the 3rd day. Further irrigations at intervals of 7-10 and 10-15 days during summer and winter season, respectively may be given depending on soil and weather conditions. Soybean is very sensitive to excess moisture and the crop is affected, if water stagnates in the fields. The crop should not suffer due to water stress from flowering to maturity. Water requirement is 300mm rainfall.To alleviate moisture stress spray of either Kaolin 3% or liquid paraffin at 1% on the foliage.

10.3 Blackgram, Greengram and Lablab

Irrigate immediately after sowing. Life irrigation on the third day. Irrigate at intervals of 7 to 10 days depending upon soil and climatic conditions. Flowering and pod formation stages are critical periods when irrigation is a must. Avoid water stagnation at all stages. Water requirement is 300mm rainfall. Apply KCl at 0.5% as foliar spray during vegetative stage if there is moisture stress.

10.4 Bengal gram

Although, most of the chickpea crop in India is grown on moisture conserved in the soil from the rain received prior to planting, the crop responds favourably to supplemental irrigation. Irrigation during the pre-flowering period and at early pod filling stage results in increasing yields. Irrigation improves the nodulation and increase the yield by increased pod number/plant. Excess moisture may be equally harmful as the crop may be affected not only by lack of aeration but also, by poor nodulation and low rhizobial activity improved cultural practices, like bunding, land leveling, adoption of proper dry farming practices, stirring the soil to improve infiltration, etc., are therefore necessary. Water requirement of crop is 100-200mm.

10.5 Cowpea

Water is applied immediately after sowing followed by life irrigation on the third day. Then, irrigation is done whenever, the fields become dry. Irrigation during flowering, pod formation and seed development are must. Water requirement is 300mm.

10.6 Drip irrigation for pulses

Drip irrigation to pulses is a recent concept where small frequent irrigation applications are applied to saturate the soil and meet the plant water requirement, while reduced leaching and run-off. Pulse drip irrigation increased grain yield by 12% compared with continuous irrigation.

QUESTIONS

I. Fill in the blanks

1. Rice is a _______________ plant
2. Water requirement of rice crop is _______ mm
3. Moisture sensitive stage of wheat is ____________
4. Water requirement of finger millet is ___________ mm
5. Drip irrigation increases the yield of pulses around _____ %

II. Choose the best answers

1. Commonly practiced irrigation method for rice is ____________
   (a) Border irrigation (b) Flood irrigation
   (c) Strip irrigation (d) Basin irrigation
2. Water loss through percolation in rice field is ___________
   (a) 40 to 70% (b) 75% (c) 30% (d) 25%
3. Method of irrigation suitable for maize crop is__________
   (a) Flood irrigation (b) Check basin
   (c) Furrow irrigation (d) None of the above

4. Moisture sensitive stages of maize are ____________
   (a) Silking (b) Tasseling (c) A & B (d) Vegetative stage
5. Moisture sensitive stages of pulses are __________ & __________
   (a) Flowering (b) Pod formation (c) Seedling (d) A & B

III. Answer the following questions

1. Give details on water requirement of rice crop at various growth stages
2. List out the methods of irrigation suitable for cereals and pulses
3. What isDASM and its importance in irrigation scheduling?
4. Explain irrigation management strategies for wheat
5. Explain irrigation management strategies for maize
6. Water management for rice, maize, pearl millet, finger millet, minor millets
7. Drip irrigation for pulses
8. What is skip furrow irrigation?
9. What is alternate furrow irrigation?

7

WATER MANAGEMENT FOR OILSEED CROPS AND COMMERCIAL CROPS

1. OILSEEDS

1.1 Sunflower

Sunflower responds to irrigation with yield increases of 100 to 200% over dryland yields common on droughty soils. Sunflower is more tolerant to drought than most crops but seed yields are higher under irrigated condition during flower development (about 7-8 weeks after emergence), rapid plant growth utilizes more moisture. To avoid moisture stress during this stage, the crop should be irrigated weekly if rain does not occur that often. Blaney-Criddle method was used in the determination of the water requirement for sunflower.

1.2 Groundnut

Suitable method of irrigation is border and strip method but farmers go for check basin method of irrigation. Usually two irrigations are required namely first irrigation at the time of sowing and second irrigation at 20-25 days after sowing. Irrigation is not necessary beyond 90 days after sowing. Hard soils are given a very light irrigation to wet 5cm depth of soil just before harvesting for pulling out the plants easily.

1.3 Castor

Pre-planting irrigation is necessary. The next irrigation is at 6 to 8 leaf stages. Heavy irrigation at long intervals are preferable after early growth period since light irrigations encourage root proliferation near the soil surface and renders the plant susceptible to drought injury. Moisture stress induce flowering. Irrigation is stopped 3 to 4 weeks before harvesting. Water requirement is about 500 mm.

1.4 Mustard

It is a rain fed crop during winter season (September-May). Due to low and uncertain rainfall during crop growing season these crops respond to irrigation. Flowering and pod or siliqua development stages are moisture sensitive stages. Irrigation may be scheduled at 50% depletion of available moisture. The IW/CPE ratios suitable for these crops vary from 0.6-0.8. The water requirements of these crops vary from 300 to 500 mm.

1.5 Sesame

Sesame is grown under irrigation in summer. First irrigation is given immediately after sowing to aid in germination and promote the growth of seedlings. The second irrigation is given when the plants are about 12cm high. Maximum water requirement is at flowering and water stress at this period reduces yield considerably. Late irrigation prolongs the growing period and increases the seed loss due to shattering. 3-6 irrigations are adequate and the water requirement is about 300mm.

2. COMMERCIAL CROPS

2.1 Cotton

In north India,pre- sowing irrigation is given followed by second irrigation after 3-4 weeks of seed germination, whereas in South India the irrigation interval is 1-2 weeks. A normal crop requires 6-8 irrigation. Irrigation must be reduced after flowering and almost stopped after first boll opened.

Critical stages

- Sympodial branching – 60-70 DAS
- Flowering - 90-100 DAS
- Boll formation stage – 125 DAS
- Boll bursting stage – 140 DAS

Irrigation methods in cotton

Cotton plant reacts strongly to moisture conditions in the soil. Too much as well as too little of soil water affects it equally adversely.

Skip furrow irrigation

- Suited to heavy soils like clay and loamy
- Alternate furrows should be skipped and may be converted to ridges having a wide bed formation.
- Short term crops like pulses may be raised in wider bed without exclusive irrigation.
- Water saving is 50% when compared to control.

Alternate furrow irrigation

- During any one run of irrigation a particular set of alternate furrows is irrigated.
- The interval of irrigation should be shortened when compared to the conventional furrows.
- During the next run, the left over furrows be irrigated.
- Suited to heavy soils like clay and loamy.

Sprinkler irrigation

- This is specially suitable to shallow sandy soils of uneven topography, where leveling is also not practicable and to the regions where both labour and water are scarce.
- Sprinklers are advantageous compared to the surface methods as water can be delivered at a desired and controlled rate, thereby ensuring a uniform distribution of water and high (water use) efficiency.
- Although more extensively used in advanced countries, it is not popular in India because of high initial investment cost.

Flexible drip system

- The Long Staple cotton tract is competing with other high value crops such as turmeric, sugarcane, tapioca and vegetables.
- Planning of flexible drip system by permanent laying out of main and sub-mains on field bunds and using laterals for cotton has a greater applicability of the system in a holistic manner.

2.2 Sugarcane

Period of irrigation

In sandy loam soils, the normal depth of irrigation under furrow method of irrigation is 2.5 to 3.75 ha-cm at the early formative phase. At grand growth including stalk elongation, the irrigation depth is 5.0 to 6.5 cm. Experiments have also demonstrated that sugarcane can be irrigated at 75% available soil moisture or 50% available soil moisture, if a trash blanket of 5-8 t /ha is provided.

Methods of irrigation

Furrow irrigation

This is the most common method of irrigating sugarcane. Initially, ridges and furrows are made and cane is planted in the furrows. At final earthing up (3-4 months later), the ridge becomes a furrow and serves as an irrigation channel. At this stage, long furrows are connected in the form of serpent; it is popularly known as serpentine method of furrow irrigation. In a level land, the furrow length could be of 50-100m. In Tamil Nadu, 'check-basin cum furrow' irrigation is followed. In this, the water is let into a set of 5-6 furrows which are connected at the both ends. However, the check basin method of irrigation is advocated in sodic soils.

Alternate furrow/skip furrow irrigation

Under water scarcity situations, it is advisable to adopt skip / alternate furrow irrigation. This planting economized the total water use by about 32%. Another low cost technology is to irrigate alternately –alternate irrigation. There is at least a 30% saving of water and this is suited to areas under well irrigation and drought-prone areas.

Surge irrigation

Surge irrigation is the regulated, pulsed application of water in furrows, irrigating the permeable soils in furrows by interrupting the irrigation when only half or two-thirds of the advance is completed .The irrigation is then completed 12-48hr. later.

Surge irrigation would restrict deep percolation losses and permit uniform water application. The water saving is 25-33% in sandy soils and 20% in loamy and clayey soils.

Drip or trickle irrigation

Drip irrigation is the frequent and slow application of water to the base of the root zone of each plant through mechanical devices or holes called emitters.

Types of drip irrigation

- Surface drip system.
- Sub-surface drip system.

Water so delivered at the root zone and there are no water losses. In subsurface system, lateral pipes are buried in the soil near the root zone and water is delivered in trickles.

Pit method of sugarcane planting under drip fertigation system

Pits are made with the spacing of 1.5 x 1.5 m with the depth of 0.38m and the diameter of 0.9m.Then they are filled with compost and native soil and mixed well. The setts are placed in a circular fashion with equal spacing and covered with soil. On 50 to 60 days, partial earthing up is given followed by full earthing-up at 90 to 100 days after planting. Drip design is lateral to lateral spacing of 3.0m (alternate rows). Irrigation is given either daily or in alternative days.

2.3 Sugarbeet

Sugar beet is a sugar producing crop in which sugar is stored in roots. Unlike sugarcane, which is the crop of tropics, sugarbeet is essentially a crop of temperate regions. Sugarbeet is a *Rabi* season crop of 6-7 months duration and is generally sown in October. The crop requires 8 to 10 irrigations depending on the seasonal conditions. Sugarbeet is very sensitive to moisture stress. Two irrigations in the formative stage, two in the leaf development stage and four in the root development stage are the minimum requirement for this crop. Excessive irrigation to the crop is highly detrimental to shoot quality.

2.4 Tobacco

A light irrigation before transplanting is essential to assure good stand. Irrigation is also essential from the time the plants are about 45-60 cm in height up to full bloom or topping stage to aid in rapid growth. Cutting of inflorescence to encourage leaf growth is known as topping. This is the period of rapid growth of the plant. Water requirements of the plants are high and leaf quality and yield are largely determined during this period. Increase in the number of leaves and enlargement in the leaf size takes place. The crop should not experience moisture stress during this period. Tobacco grown in light soils and in *Rabi* season requires irrigation.

When flue-cured tobacco is raised in black soils as a rainfed crop, one irrigation is essential 40 to 50 days after planting. This irrigation was found to increase yields without affecting quality. This is the period when water demand of the crop is high and the black soil starts cracking due to low moisture level. About 60 to 75 cm depth of soil is crucial both for root activity and moisture supply. The chloride content of irrigation water should be less than 40 ppm. Irrigation at later stage delays maturity.

Flue-cured tobacco can raised successfully in light soils if adequate and timely irrigation is provided. Excessive irrigation leaches the nutrients from the soil and slick leaf with dirty colour is produced. Inadequate irrigation retards crop growth and curing becomes difficult. After establishment of the crop, moisture stress is allowed for two weeks followed by first irrigation. The second irrigation is given when the crop is at knee high stage. Two to three irrigations may be given between knee high stage and topping. After topping, moisture stress is allowed for two weeks. Afterwards, one or two irrigations may be given depending on the curability of the leaf. Each irrigation is of 40 mm. Tobacco is highly sensitive to excess moisture. The water requirement of tobacco ranges from 300 to 400 mm.

QUESTIONS

I. Fill in the blanks

1. Sunflower responds to irrigation with yield increases of 100 to ______%over dryland.
2. Water requirement of castor is about ________ mm
3. Maximum water requirement stage of sesame is _____________
4. Critical stage for irrigation in sugarcane is ____________ stage
5. Sugarbeet requires ___________ irrigations.

II. Choose the best answers

1. Suitable method of irrigation for groundnut
   (a) Border irrigation (b) A & C
   (c) Strip irrigation (d) Basin irrigation

2. ____________ and ___________ are moisture sensitive stages of mustard
   (a) Flowering & siliqua (b) Germination & flowering
   (c) Branching & flowering (d) Germination & branching
3. Cotton requires ________ numbers of irrigation
   (a) 3 - 4 (b) 7 - 8 (c) 5 - 6 (d) 6 - 8
4. Excessive irrigation to __________ is highly detrimental to shoot quality
   (a) Sugarbeet (b) Sugarcane (c) Tobacco (d) Sunhemp

III. Answer the following questions

1. Give details about method of irrigations followed in cotton.
2. Explain pit method of sugarcane planting

8

WQUALITY OF IRRIGATION WATER - CLASSIFICATION OF QUALITY OF IRRIGATION WATER BASED ON DIFFERENT SALTS- AGRONOMIC PRACTICES FOR USE OF POOR QUALITY WATER

1. QUALITY OF IRRIGATION WATER

Irrigation water drawn from different sources *viz,* river, canal, tank, open well or tube well always contain some soluble salts dissolved in it. Water being a universal solvent, several salts are dissolved in irrigation water. The quality and quantity of salt present in the water depend on the nature of water sources and the soils and underground strata over which the water flows. The main soluble salts in water are Ca, Mg, Na and K as cations and chloride, sulphate, carbonate and bicarbonate as anions. The amount and kind of salts present determines the suitability of water for irrigation.

2. CLASSIFICATION OF QUALITY OF IRRIGATION WATER BASED ON DIFFERENT SALTS

Classification and suitability of irrigation water for use to crops are based on the following criteria.

1. Total salt concentration
2. Relative proportion of sodium to other cations
3. Carbonate and bicarbonate concentration and
4. Boron concentration

2.1 Total salt concentration in irrigation water

Total concentration of soluble salts is the most important single criterion of irrigation water quality. The harmful effects increase with increase in total salt concentration.

The electrical conductivity (EC) has now been accepted as the standard method of evaluating irrigation water for total salts because of its reliability and ease of determination. Electrical conductivity is the reciprocal of the electrical resistance and is now expressed as desi Siemens per metre (dSm^{-1}), which is equal to one milli-mhos cm^{-1}. EC serves as a guide to know the extent of soluble salts in irrigation water.

Soil Testing Laboratory of Tamil Nadu classified irrigation water based on EC as follows.

Quality of water	EC (dSm^{-1})	Suitability
Good	< 1	Suitable for irrigation
Critical	1 – 3	Suitable on proper management
Injurious	> 3	Not suitable

2.2 Relative proportion of sodium to other cations in irrigation water

The presence of sodium salts and their concentrations in water decide largely the quality of water for irrigation. The Sodium Adsorption Ratio (SAR), Exchangeable Sodium Percentage (ESP) and Sodium Percentage (SP) are considered for evaluating sodium problems in water. Water may contain carbonate and bicarbonates that aggravate the sodium hazards by precipitating the calcium and magnesium ions.

2.2.1 Sodium Adsorption Ratio (SAR)

It measures the sodium cations in relation to the calcium and magnesium ions present in a solution.

$$SAR = \frac{Na}{\sqrt{Ca + Mg/2}}$$

Table 10. Classification of irrigation water according to sodium hazards

Sodium hazards	Class	SAR	Suitability
Low	S_1	0 – 10	Can be used on all soil with little danger of accumulation of exchangeable sodium
Medium	S_2	10 – 18	Can be used in all soils with good permeability; appreciable sodium hazards in soils of high clay and low organic matter.
High	S_3	18 – 26	Tends to cause harmful sodium accumulation in most soils: needs good drainage, high leaching and organic matter addition.
Very high	S_4	> 26	Generally unsatisfactory for irrigation

2.2.2 Exchangeable sodium percentage (ESP)

Exchangeable sodium percentage is closely related to sodium adsorption ratio and gives the degree of saturation of the soil / water exchange complex with sodium.

$$ESP = \frac{\text{Exchangeable sodium}}{\text{Cation exchange capacity}} \times 10$$

2.2.3 Sodium Percentage

(SP) refers to proportion of sodium in relation to total cations in water and the sodium percentage is calculated as follows.

The exchangeable cations on soil colloids are in equilibrium with sodium ions in soil solution. As the sodium percentage in irrigation water increases, there is a tendency to greater adsorption of sodium ions on the clay complex leading to alkalization of the soil. If the SP is more than 60, the use of water for irrigation is questionable.

$$SP = \frac{\text{Na}}{\text{Na} + \text{Ca} + \text{Mg} + \text{K}} \times 100$$

2.3 Carbonate and bicarbonate contents

Carbonates and bicarbonates have an indirect influence on the water quality. Calcium and magnesium reacts with carbonates and bicarbonates in irrigation water and precipitates as insoluble carbonates. Loss of calcium and magnesium from water increases sodium percentage causing sodium hazards. This precipitation is favoured by drying out of soils containing solution high in bicarbonates and calcium and magnesium ions.

The term Residual Sodium Carbonate is used to indicate the relative proportion of carbonates and bicarbonates in relation to calcium and magnesium in water.

The RSC is expressed as RSC = $(CO_3^{2-} + HCO_3^-) - (Ca^{2+} + Mg^{2+})$

Suitability of water for irrigation based on RSC is classified as

RSC (me/l)	Suitability of water for irrigation
< 1.25	Good
1.25 - 2.50	Fair
2.50	Bad

When appreciable RSC occurs more frequent irrigations is necessary to prevent soil drying and to do leaching at frequent intervals.

2.4 Boron concentration

Boron is essential to plants and it helps growth when present in very small amount in irrigation water. Boron toxicity occurs when it is present in soil solution in more than 3.0 ppm.

3. AGRONOMIC PRACTICES FOR USE OF POOR QUALITY WATER

3.1 Saline water management

Whenever, it is inevitable to use water of poor quality for crop production appropriate management practices helps to obtain reasonable yield of crops. Some of the important management practices are as follows:

(a) **Application of gypsum:** Chemical amendments such as gypsum, when added to water will increase the calcium concentration in the water, thus reducing the sodium to calcium ratio and the SAR, thus improving the infiltration rate. Gypsum requirement is calculated based on relative concentration of Na, Mg & Ca ions in irrigation water and the solubility of gypsum. To add 1 meq/L of calcium, 860 kg of gypsum of 100% purity per ha m of water is necessary.

(b) **Alternate irrigation strategy:** Some crops are susceptible to salinity at germination & establishment stage, but tolerant at later stage. If susceptible stages are ensured with good quality water, subsequent tolerant stages can be irrigated with poor quality saline water.

(c) **Fertilizer application:**Application of fertilizers in small doses and frequently improve uptake and reduce damage to the crop plants. In addition, the lower the salt index of fertilizer, the less danger there is of salt burn and damage to seedlings or young plants.

(c) **Methods of irrigation:**Poor quality irrigation water is not suitable for use in sprinkler method of irrigation. Crops sprinkled with waters having excess quantities of specific ions such as Na and Cl cause leaf burn. High frequency irrigation in small amounts as in drip irrigation improves water availability and uptake due to micro leaching effect in the wetted zone.

(e) **Crop tolerance:**The crops differ in their tolerance to poor quality waters. Growing tolerant crops when poor quality water is used for irrigation helps to obtain reasonable crops yields.

Degree of salt tolerance: Good- Turnips, Beet root, Date palm. Moderate - Lettuce, Tomato, Asparagus, Carrot, Spinach, Onion, Pepper, Clove, Fig, Grape, Olive. Poor - Peas, Celery, Cabbage, Egg plant, Sweet potato, Potato, Green beans, Plum, Pear, Apple, Orange,

(f) **Method of sowing:**Placing the seed on the slope of the ridge, several cm below the crown is recommended for successful crop establishment.

(g) **Drainage:** Provide adequate internal drainage. Meet the necessary leaching requirement depending on crop and EC of water. This is necessary to avoid build of salt in the soil solution to levels that will limit crop yields. Leaching requirement can be calculated from water test results and tolerance levels of specific crops.

(h) **Other management practices**

(i) **Over aged seedlings in rice:** Closer spacing of over aged seedlings results in better establishment in salt effected soils than normal aged seedlings.

(ii) **Mulching:** Mulching with locally available plant material help in reducing salt problems by reducing evaporation and by increasing infiltration.

(iii) **Soil management:** All soil management practices that improve infiltration rate and maintain favourable soil structure reduces salinity hazard.

(iv) **Crop rotation:** Inclusion of crops such as rice in the rotation reduces salinity.

(v) **Conjunctive use of irrigation water:** Conjunctive use of water relates to the combined use of ground water or poor quality water and *surface water*. Due to the augmented water source, higher water reliability can be achieved. Conjunctive use therefore functions as a buffer for periods of water scarcity or act as a supplement where poor quality water is used for irrigation. The idea of this management approach is to use *surface* *water*when the *water table*is high and change to *groundwater* when the *water table*is low. This technique might be especially important as a buffer function for mitigating impacts of climate change (such as increased heat, drought) and important management practices where saline or poor quality water is used for irrigation.

3.2 Effluent water management

(a) Industrial waste utilization for land reclamation and crop production

- Untreated distillery effluent (Spent wash) to reclaim and improve the productivity of the sodic soils is now advocated. Untreated distillery effluent (spent wash) is acidic (pH: 3.8 – 4.2) with considerable quantity of potassium. Calcium and magnesium and traces of micronutrients.
- 3.75 to 5.00 lakhs litres of untreated distillery effluent (spent wash) per hectare of sodic soils in summer months is recommended.

(b) Irrigation of pulp and paper mill effluents

- Pulp and paper effluents contain lot of dissolved solids and stabilized organic matter. The properly treated effluent with EC less than 1.2 dSm^{-1} as such can safely be used for irrigation with appropriate amendments viz., pressmud @ 5 tonnes ha^{-1}.
- Tolerant crops such as rice, maize, sunflower, groundnut, soybean, sugarcane and tapioca.

(c) Treated tannery effluent irrigation

- Leather industry is one the major foreign exchange earners in India.
- Tolerant crops such as rice, ragi, sunflower, sugarcane, tomato can be grown.

3.3 Use of sewage water for irrigation

The rapid population growth in many municipalities in the arid and semiarid of world continues to place increased demands on limited fresh water supplies. Many cities and districts are struggling to balance water use among municipal, industrial, agricultural, and recreational users. The population increase has not only increased the fresh water demand but also increased the volume of waste water generated. Treated or recycled waste water (RWW) appears to be the only water resource that is increasing as other sources are dwindling.

The use of sewage water for irrigation is probably as ancient as humankind's cultivation of land. Large-scale, controlled sewage water use for irrigation, however, only dates back to last century when so-called sewage farms were established in parts of Europe, Australia, India and the United States for the purpose of disposing of waste water and preventing river pollution. Although agricultural crops were produced on these farms, crop production was a secondary consideration. During this century, and particularly within the past two decades, the practice of irrigating agricultural crops with municipal waste water has become more widespread, especially in arid and semiarid areas of both developed and developing countries. The controlled use of untreated and treated waste water in irrigation is now quite common in Europe, the United States, Mexico, Australia, China, India and the Near East as well as, to a lesser extent, in Chile, Peru, Argentina, the Sudan and South Africa.

Sewage water

Sewage indicates the liquid waste from the community. It includes sludge, discharge from latrines, urinals, stable industrial waste and also the ground surface and storm water that may be admitted into the sewer.

Benefits of using sewage water for irrigation

There are many potential benefits offered by the use of municipal waste water for irrigation purposes, including the safe and low-cost treatment and disposal of waste water; the conservation of water and recharge of groundwater reserves; and the use of nutrients in the waste water for productive purposes. The irrigation of trees may provide additional benefits. In areas where water is relatively plentiful, crop or forestry irrigation may be adopted primarily for treatment and disposal purposes. In arid and semi-arid zones, the issues of groundwater recharge and tree and crop production may be equally or more important.

- Safe, low-cost treatment and disposal
- Improved use and conservation
- Nutrients for crop production
- Greening cities
- Municipal waste water in forestry irrigation

QUESTIONS

I. Fill in the blanks

1. The amount and kind of ______ present in the water determines the suitability of water for irrigation
2. Total concentration of ___________ is the most important single criterion of irrigation water quality
3. Unit of electrical conductivity is ___________
4. Boron toxicity occurs when soil solution has more than _____ ppm boran
6. To add 1 meq/L of calcium, _______ kg of gypsum per ha m of water is necessary
7. ______________ irrigation strategy can be used for crops susceptible for salt concentration at germination stage.

II. Choose the best answers

1. ___________used to evaluate the total concentration of salts in irrigationwater

   (a) EC (b) pH

   (c) OH (d) None of the above

2. EC of good quality water is ____________

   (a) 1-2 dS/m (b) < 1 dS/m (c) 3 dS/m (d) > 3 dS/m

3. ___________ me/l of RSC of water highly suitable for irrigation

   (a) < 1.25 (b) 1.25 – 2.50 (c) > 3.50 (d) > All the abobe

4. ___________ application increase the infiltration rate of soil

   (a) Fertilizer (b) Gypsum

   (c) Rockphosphate (d) None of these

III. Answer the following questions

1. Define quality of irrigation water
2. List out the salts present in the water
3. List out the criteria used for classifying the quality of irrigation water
4. Give the classification of irrigation water according to Na hazards
5. List out the crops tolerant to salt concentration
6. Classification of quality of irrigation water based on different salts
7. Agronomic management for use of poor quality water
8. Effluent water management in Agriculture

9

MEASUREMENT OF IRRIGATION WATER

Water is the most valuable commodity in irrigated agriculture. Accurate measurement of irrigation water provides efficient use of this scarce natural resource.

1. UNITS OF MEASUREMENT

Water at rest (in reservoirs, ponds, soil, tanks) can be measured in units of volume such as litres, cubic metre, hectare centimetre and hectare metre.

Water in motion (rivers, canals, pipeline, field channels and channel structures) is measured in rate of flow units such as litres per hour and hectare metre per day.

Litre	A volume equal to one cubic decimeter or 1/1000 cubic metre
Cubic metre	Cubic metre equals to 1000 litres
Hectare centimetre	A volume necessary to cover an area of 1 hectare (10000m^2) to a depth of 1cm = 10000 m^2 × 0.01 m = 10 0m^3 = 1,00,000 litres
Hectare metre	A volume necessary to cover an area of 1 hectare to a depth of 1 metre = 10000 m^2 × 1 m = 10000 m^3 = 10 million litres.
Litre/second	Continuous flow amounting 1 litre pass through a point for each second
Cubic metre / second	A flow of water equivalent to a stream of 1 metre wide and 1 m deep flowing at a velocity of 1 metre per second

2. METHODS OF WATER MEASUREMENT

I. Volumetric measurements

II. Velocity-area methods

III. Direct discharge methods (orifices, weirs, flumes)

IV. Tracer methods (dilution)

2.1 Volume method

Materials required

Plastic bucket, alkathene pipe and stop watch.

Procedure

- A graduated bucket is taken and placed near to the downstream.
- The alkathene pipe is used to fill the bucket.
- The time taken to fill the bucket is recorded using the stop watch. The rate of flow of the stream is calculated by the formula.

  $$\text{Discharge (lit/sec)} = \frac{\text{Volume of bucket (litre)}}{\text{Time taken to fill the bucket (second)}}$$

- Graduated bucket will be useful for smaller stream size like water taps.
- For larger stream sizes like water discharged from water tanks, pump set, the discharge rate can be worked out with the volume of the container and the time taken to fill it.

2.2 Velocity area method

- This method is used to determine the discharge rate of water in a pipe or open channel by multiplying the cross sectional area of flow section at right angles to the direction of flow by the average velocity of water.

  Discharge = Area × Velocity

  $Q = a \times v$

  Where,

  Q - Discharge rate (m^3/sec)

  a - Area of cross section of channel (m^2)

  v - Velocity of flow (m/sec)
- The cross sectional area is determined by direct measurements. The velocity is measured by a current meter.

2.2.1 Float method

Principle: It involves recording of the rate of movement of a floating body in the running water course.

Materials: A long necked bottle, a block of wood or any floating material, tape, stop watch.

Procedure

- Measure 30 m length in a straight channel and mark the upstream and down stream points.
- Allow the wooden float or the bottle partly filled with water to float on the running water well above the upstream point.
- Note the time when it touches the upstream point and let this be the initial time.
- Then note the time when it reaches the downstream point which will be the final time.

- Repeat the procedure several times and find out the average time taken for the float to cross the 30 m distance.
- The velocity is determined by the following relationship.

$$\text{Measured Velocity} = \frac{\text{Length of channel}}{\text{Average time taken by float}}$$

- From this, the average velocity is calculated by multiplying a coefficient factor.

  Average velocity = Measured velocity x 0.85 (coefficient)
- The flow rate of water is worked out using the formula

  Rate of flow = Average velocity × Cross sectional area of the channel

2.2.2 Water meter method

- Water meter has a multi-blade propeller made of metal, plastic or rubber, rotating in a vertical plane and geared to a totalizer in such a manner that a numerical counter can totalize the flow in any desired volumetric units.
- The basic requirements are (1) the pipe must flow full at all times; (2) the rate of flow must exceed the minimum range.
- When water meters are installed in open channels, the flow must be brought through a pipe of known cross sectional area.
- Water meters are very costly.

2.3 Direct discharge methods

In this method, the volume of flow of water is determined directly by installing certain devices of known dimensions at a desired point across the channel. Devices for measuring the irrigation water are, Weirs, flumes, orifices and meter gates.

2.3.1 Weirs

- Weirs are used to measure the flow in an irrigation channel or the discharge of a well or canal outlet at the source.
- A weir is a notch of regular form through which the irrigation stream is made to flow.
- Weirs may be built as stationary structures or they may be made portable.
- Rectangular weirs and 90°V notch weirs are commonly used.
- The discharge through a weir is calculated as

  $Q = CLH^m$

Where,

Q - Discharge

C - Coefficient

L - Length of crest

H - Head of the crest

m - Exponent depending on weir opening

2.3.1.1 Triangular weir (90°V notch)

- It is commonly used to measure small and medium size streams accurately.
- The discharge through 90°V notch weir may be computed either by using the following formula or standard table (Table 1).

  $Q = 0.0138\ H^{5/2}$

Where,

Q - Discharge (lit/sec)

H - Head (cm)

Table 1. Discharge rate through 90°V-notch (lit/sec)

Height of water over V-notch (cm)	Discharge (lit/sec)	Height of water over V-notch (cm)	Discharge (lit/sec)	Height of water over V-notch (cm)	Discharge (lit/sec)
1	2	1	2	1	2
5.0	0.8	14.0	10.5	23.0	35.7
6.0	1.2	15.0	12.3	24.0	40.0
7.0	1.8	16.0	14.5	25.0	44.5
8.0	2.5	17.0	16.7	26.0	48.8
9.0	3.4	18.0	19.4	27.0	53.8
10.0	4.5	19.0	22.3	28.0	58.7
11.0	5.7	20.0	25.5	29.0	64.5
12.0	7.1	21.0	28.3	30.0	69.4

2.3.1.2 Rectangular weir

- It is used to measure comparatively larger discharges.
- It has a horizontal crest and vertical sides.
- They may be either contracted rectangular weirs or suppressed rectangular weirs.
- The discharge through rectangular weirs may be computed by the Francis formula stated below
  - (i) Suppressed rectangular weir:

    $Q = 0.0184\ LH^{3/2}$
  - (ii) Contracted rectangular weir (with end contraction at one end):

    $Q = 0.0184(L-0.1H)H^{3/2}$
  - (iii) Contracted rectangular weir (with end contraction at both ends):

    $Q = 0.0184(L-0.2H)H^{3/2}$

Where,

Q - Discharge

L - Length of crest (cm)
H - Head over the weir (cm)

2.3.1.3 Trapezoidal weir (Cipoletti weir)

- This is invented by an Italian Engineer by name 'Cipoletti' therefore it is called as Cipoletti weir.
- Each side of the weir has a slope of 1 horizontal to 4 vertical. It is used to measure medium discharges.
- The discharge through cipoletti weir is computed by the following formula

  $Q = 0.0186\ LH^{3/2}$

Where,

Q - Discharge
L - Length of crest (cm)
H - Head over the weir (cm)

2.3.2 Flumes

2.3.2.1 Parshall flume (Venturi flume)

- This has been designed by Parshall (1950) and hence named after him.
- Parshall flume is an open channel device that operates with a small drop in head.
- It is a self cleaning device.
- Sand or silt in the flowing water does not affect its operation or accuracy.
- It gives reasonably accurate measurement even when partially submerged.
- Flumes of 7.5, 15, 25 and 30 cm sizes are generally used in field measurement.
- The discharge rate can be calculated by using the standard table (Table 2).

Table 2. Free-flow discharge values for Parshall Flume (lit/sec)

Head (cm)	Throat width			
	7.5 cm	15 cm	23 cm	30 cm
4	1.2	2.3	4.0	4.5
5	1.7	3.3	5.5	7.0
6	2.3	4.4	7.2	9.6
7	2.7	5.4	8.5	11.4
8	3.5	7.2	11.1	14.4
9	4.3	8.5	13.5	17.7
10	5.0	10.2	15.9	21.1
11	5.8	11.6	18.1	23.8
12	6.7	13.5	21.1	27.5
13	7.5	15.0	23.3	31.0

14	8.5	17.3	26.7	35.0
15	9.4	19.2	29.5	38.7
16	10.4	21.2	32.5	42.7
17	11.4	23.2	35.6	46.7
18	12.4	25.3	39.0	51.2
19	13.6	27.8	42.5	55.0

2.3.2.2 Cut throat flume

- It is an improvement over Parshall flume.
- This flume has a flat bottom, vertical walls and zero length throat section i.e. it has no throat section.
- The flume length may vary from 45 cm to 3.0 m and the width ranged between 2.5 cm to 1.8 m.
- Cut throat flume can operate either in free flow or in submerged condition.
- It should be installed in a straight section of the channel and not near the gate because of unstable and surging effects which might result from the gate operation.
- It is better to have a flow measuring device operate under free flow condition.
- The advantage is its economy.

2.3.3 Orifices

- Orifices are circular or rectangular openings in a vertical bulkhead through which water flows.
- They may be operated under free flow or submerged flow conditions.
- Under free flow conditions the flow from the orifice discharges entirely into the air.
- In submerged flow orifices, the downstream water level is above the top of the opening and the flow discharges into water.
- Free flow orifice plates can be used to measure comparatively small streams like the flow into border strips, furrows or check basins.
- In submerged flow orifice, plastic scale may be fixed on the upstream face of the orifice plate such that the zero of the plastic scale coincides with the centre of the orifice.
- The discharge through an orifice is calculated by the following formula

  $Q = 0.61 \times 10^{-3}\ a \sqrt{2gh}$

Where,

Q - Discharge (lit/sec)

a - Area of cross section of the orifice (cm^2)

g - Acceleration due to gravity (cm/sec^2) (981 cm/sec^2)

H - Depth of water over the centre of the orifice (cm)

Submerged orifices

(i) Those having orifices of fixed dimensions

(ii) Those in which the height of the opening may be varied

2.3.4 Meter gate

- A meter gate is basically a modified submerged orifice so arranged that the orifice is adjustable in area.
- They are used to control the water flowing from one channel to another.
- They are used to measure the rate of flow if the head and area of opening can be determined and the gate has been calibrated.

2.4 Tracer method (dilution method)

A substance (tracer) in concentrated form is introduced into the flowing water and allowed to thoroughly mix. The concentration of the tracer is measured in the downstream section. Only the quantity of water necessary to accomplish the dilution is involved, there is no need to measure the velocity, depth, head, cross sectional area or any other hydraulic factors usually considered in discharge measurements.

QUESTIONS

I. Fill in the blanks

1. Water discharge rate is calculated in volumetric method by using formula____________
2. Formula used in velocity area method is ________________.
3. Formula used to compute discharge rate of water for suppressed rectangular weirs __________________.
4. Parshall flume designed by __________ during ________.

II. Choose the best answer

1. Formula used to calculate the discharge rate by using weir is ________

   (a) Q = CLHm (b) Q = 0.0138 H 5/2

   (c) Q = 0.0184 LH 3/2 (d) Q = 0.0184(L-0.1H)H 3/2

2. Formula used to calculate the discharge rate by using Triangular weir (90°V notch) is ________

   (a) Q = CLHm (b) Q = 0.0138 H 5/2

   (c) Q = 0.0184(L-0.1H)H 3/2 (d) Q = 0.0184 LH 3/2

3. The discharge through contracted rectangular weir (with end contraction at one end) computed by formula

   (a) Q = CLHm (b) Q = 0.0138 H 5/2

   (c) Q = 0.0184(L-0.1H)H 3/2 (*d*) Q = 0.0184 LH 3/2

4. The discharge through an orifice is calculated byformula

   (a) Q = CLHm (b) Q = 0.0138 H 5/2

   (c) $Q = 0.61 \times 10^{-3}$ a √ 2gh (d) Q = 0.0184 LH 3/2

5. __________ is basically a modified submerged orifice

   (a) Water meter (b) Flumes (c) Meter gate (d) All the above

III. Answer the following question

1. What is water meter?
2. List out devices used in direct discharge method of water measurement
3. Give details on fumes and orifices
4. List out different weirs used for water measurement with formula.

10

SCHEDULING OF IRRIGATION

Irrigation scheduling is the application of optimum quantity of water at optimum time intervals for obtaining potential crop yield. It is the decision making process indicating when irrigation water is to be applied and the quantity of water to be applied each time.

1. IRRIGATION SCHEDULING

Irrigation scheduling may be defined as scientific management technique of allocating irrigation water based on the individual crop water requirement (Etc) under different soil and climatic condition, with an aim to achieve maximum crop production per unit of water applied over a unit area in unit time.

2. OBJECTIVES OF IRRIGATION SCHEDULING

The main objectives of efficient irrigation scheduling are

1. High yield of good quality
2. High water use efficiency
3. Least damage to soil productivity
4. Low irrigation cost

3. MAJOR ASPECTS OF IRRIGATION SCHEDULING

Providing knowledge on correct time (When to irrigate) and optimum quantity (How much to irrigate) of water application at each irrigation to optimize crop yields with maximum water use efficiency and at the same time ensuring minimum damage to the soil properties.

4. IMPORTANCE OF IRRIGATION SCHEDULING

Irrigation scheduling has its own meaning and importance according to the nature of the work

4.1 For irrigation Engineers

Irrigation scheduling is important to cover more area with available quantity of water or to satisfy the whole command from head to tail reach in the canal or river system.

4.2 For soil scientists

It is important that the field should not be over irrigated or under irrigated, as both will spoil the chemical and physical equilibrium of the soil.

4.3 For Agronomists

It is important to get higher yield per unit quantity of water in normal situations and to protect the crop to get as much as possible yield under drought situation by means on supplying water in optimum ratio and minimizing all field losses.

5. PRACTICAL CONSIDERATION IN IRRIGATION SCHEDULING

Before scheduling irrigation in a farm or field or a command, the following criteria should be taken care for efficient scheduling.

Crop factors

(a) Sensitiveness to water shortage
(b) Critical stages of the crop and rooting depth
(c) Economic value of the crop

Water delivery system

(a) Canal irrigation or tank irrigation (It is a public distribution system where scheduling is arranged based on the decision made by public based on the resource availability).
(b) Well irrigation (individual decision is final)

Types of soil

(a) Sandy – needs short frequency of irrigation and less quantity of water
(b) Clay – needs long frequency of irrigation and more quantity of water

Salinity hazard

To maintain favourable salt balance, excess water application may be required rather than ET requirement of the crop to leach the excess salt through deep percolation.

Irrigation methods

(a) Basin method allows more infiltration through more wetting surface which in turn needs more water and long interval in irrigation frequency.
(b) Furrow method allows less infiltration due to less wetting surface which needs less water and short interval in irrigation frequency.
(c) Sprinkler method needs less water and more frequency.
(d) Drip method needs less water and more frequency.

Irrigation interval

The extension of irrigation interval does not always save water. The interval has to be optimized based on the agro-climatic situation.

Minimum spreadable depth

Cannot reduce the depth based on the water requirement of the crop alone. The depth should be fixed based on the soil type, rooting nature of the crop and irrigation method followed. The minimum depth should be so as to achieve uniformity of application and to get uniform distribution over the entire field.

6. ADVANTAGES OF IRRIGATION SCHEDULING

- It enables the farmer to schedule water rotation among the various fields to minimize crop water stress and maximize yields.
- It reduces the farmer's cost of water and labour through fewer irrigations, thereby making maximum use of soil moisture storage.
- It lowers fertilizer costs by holding surface runoff and deep percolation (leaching) to a minimum.
- It increases net returns by increasing crop yields and crop quality.
- It minimizes water-logging problems by reducing the drainage requirements.
- It assists in controlling root zone salinity problems through controlled leaching.
- It results in additional returns by using the 'saved' water to irrigate non-cash crops that otherwise would not be irrigated during water-short periods.

7. CRITERIA (APPROACHES) FOR SCHEDULING IRRIGATION

With the advancement of knowledge in the field of soil-plant-atmospheric system, several criteria for scheduling irrigations are now available and are being used by investigators and farmers. All the available criteria can be broadly classified into the following three categories:

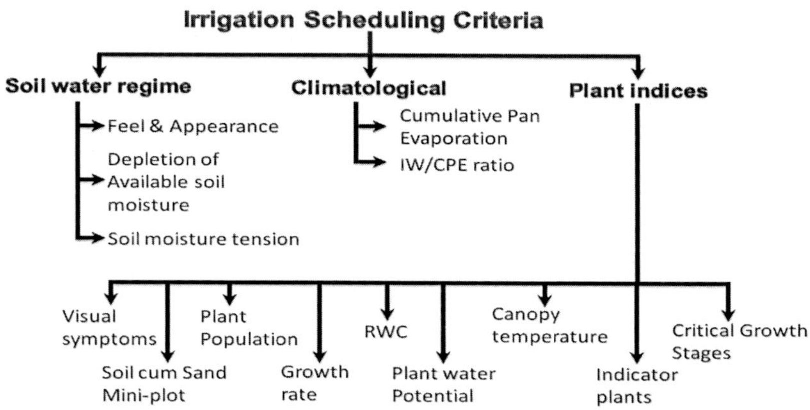

7.1 Soil water regime approach

In this approach, the available soil water held between field capacity and permanent wilting point in the effective crop root zone depth is taken as an index or guide for determining practical irrigation schedules. Alternatively, soil moisture tension, the force with which the water is held around the soil particles is also sometimes used as a guide for timing irrigations.

7.1.1 Feel and appearance of soil

This is one of the oldest and simple methods of determining the soil moisture content. It is done by visual observation and feel of the soil by hand. The accuracy of judgment improves with experience. Based on several years of experience guidelines have been developed (Table 1) which help the farmers to judge the soil moisture present in the soil samples drawn from the crop root zone depth and based on depletion of available soil moisture (DASM)

Table 1. Guide for judging the amount of available moisture in soil

Available soil moisture range	Coarse texture (Loamy sand)	Moderately coarse texture (Sandy loamy)	Medium texture (Loamy and silt loamy)	Fine texture (Clay loamy and Silty clay loamy)
Field capacity (100%)	On squeezing, no free water appears on soil, but wet outline is left on hand	Similar symptoms		
75 to 100%	Tends to stick together slightly, sometimes forms a very weak ball under pressure	Forms weak ball, breaks easily, don't slick	Forms a ball, very pliable, slicks readily	Easily ribbons out between fingers, has slick feeling
50 to 75%	Appears to be dry don't form a ball with pressure	Tends to form a ball under pressure but seldom holds together	Forms a ball somewhat plastic, some-times slick slightly with pressure	Forms a ball, ribbons out between thumb and fore-finger
25 to 50%	As above, but ball is formed by squeezing very firmly	Appears to be dry, don't form a ball unless squeezed very firmly	Somewhat crumbly but holds together with pressure	Somewhat pliable, forms a ball under pressure
0 to 25%	Dry, loose, single grained flows through fingers	Dry, loose, flows through fingers	Powdery dry, sometimes slightly crusted but easily broken down into powdery conditions.	Hard, baked, cracked, sometimes has loose crumbs on surface.

7.1.2 Depletion of the available soil moisture (DASM)

In this method, the permissible depletion level of available soil moisture in the effective crop root zone depth is commonly taken as an index or guide for scheduling irrigations to field crops. In general, for many crops scheduling irrigation at 20-25% DASM (75-80% ASM) in the soil profile was found to be optimum at moisture sensitive stages. While at other stages irrigations, scheduled at 50% DASM (50% ASM) were found optimum.

7.1.3 Soil moisture tension

Soil moisture tension, a physical property of film water in soil, as monitored by tensiometers at a specified depth in the crop root zone could also be used as an index for scheduling irrigations to field crops. Pertinent water storage properties of soil and water use rated as affected by climatic conditions and plant vigour are intrinsically accounted for in such an irrigation schedule programme without measuring them. Tensiometers are installed in pairs, one in the maximum rooting depth and the other below this zone. Whenever critical soil moisture tension is reached say for example 0.4 or 0.6 or 0.75 bars etc. in the upper tensiometer the irrigation is commenced. While the lower one (tensiometer) is used to terminate the irrigations based on the suction readings in the below soil profile zone. It is generally used for irrigating orchards and vegetables in coarse textured soils because most of the available soil moisture is held at lower tensions.

7.2 Climatological approach

The potential rate of water loss from a crop is primarily a function of evaporative demand of the atmosphere under adequate soil water conditions. Thus, in this method, the water loss expressed in terms of either potential evapotranspiration (PET) or cumulative pan evaporation (CPE) over short periods of time are taken as an index for scheduling irrigation.

Lysimeter

By isolating the crop root zone from its environment and controlling the processes that are difficult to measure, the different terms in the soil water balance equation can be determined with greater accuracy. This is done in lysimeters where the crop grows in isolated tanks filled with either disturbed or undisturbed soil

Different types of lysimeters vary widely in the accuracy with which changes in soil water are detected. In precision weighing lysimeters, where the water loss is directly measured by the change of mass, evapotranspiration can be obtained with an accuracy of a few hundredths of a millimetre and small time periods such as an hour can be considered. In non-weighing lysimeters the evapotranspiration for a given time period is determined by deducting the drainage water, collected at the bottom of the lysimeters, from the total water input. A requirement of lysimeters is that the vegetation both inside and immediately outside of the lysimeter be perfectly matched (same height and leaf area index). Lysimeters provide the direct estimates of water balance components viz., E, T, ET, drainage, runoff, effective rainfall etc in a cropped field against which other methods can be tested and calibrated. As lysimeters are difficult and expensive to construct and as their operation and maintenance require special care, their use is limited to specific research purposes.

7.2.1 Cumulative pan evaporation

Earlier investigations have shown that transpiration of a crop is closely related to free water evaporation from an open pan evaporimeter. Thus, the open pan evaporimeter being simple and as they incorporate the effects of all climatic parameters into a single entity i.e., pan evaporation could be used as a guide for scheduling irrigation's to crops. For example,

- Wheat required 75 to 100 mm CPE
- Sugarcane required 75 mm CPE
- Greengram required 180 mm CPE
- Sunflower required 60 mm CPE

7.2.2 IW/CPE ratio

The amount of water lost through evapotranspiration (ET) is estimated from climatological data and when ET reaches a particular level, irrigation is scheduled.

The most widely used approach for scheduling irrigation based on climatological data is termed as IW/CPE ratio method.

In IW/CPE ratio approach, a known amount of irrigation water (IW) is applied when the Cumulative Pan Evaporation (CPE) from an USWB open pan evaporimeter reaches a pre-determined level.

The ratio can be expressed as

$$\frac{\text{IW}}{\text{CPE}}\text{ ratio} = \frac{\text{Depth of irrigation water}}{\text{Cumulative pan evaporation}}$$

The irrigation depth (IW) for different crops are fixed based on soil, climate and root zone depth and it ranges from 4 to 6 cm, the most common being 5 cm of irrigation

For fixing the ratio of IW/CPE for each crop, experiments are conducted by irrigating the crop at different ratios and the ratio, which gives the maximum yield, is taken as the optimum IW/CPE ratio. The ratio ranges from 0.4 to 1.0 for different crops.

For arriving the cumulative pan evaporation value, the daily pan evaporation data are added till it is equal to the ratio of the amount of water applied as irrigation

The irrigation depth (IW) divided by the ration (R) will give the cumulative pan evaporation value at which irrigation is to be made

$$\text{CPE} = \frac{\text{IW}}{R}$$

For example, if the irrigation depth needed is 50 mm and the optimum ratio is 0.5, the cumulative pan evaporation value needed to irrigate the field is

$$\text{CPE} = \frac{\text{IW}}{R} = \frac{50}{0.5} = 100\text{ mm}$$

If the 100 mm CPE is attained in 10 days @ 10 mm per day of pan evaporation, irrigation is to be scheduled at 10 days interval.

Problem

Calculate the cumulative pan evaporation value at which irrigation is to be scheduled for maize, if the IW/CPE ratio recommended is 0.6 and irrigation water to be appliedis 50 mm.

$$\text{CPE} = \frac{\text{IW}}{R} = \frac{50}{0.6} = 83.3\text{ or }83\text{ mm}$$

This approach provides best correlation since climatic and soil parameters are considered.For use by farmers, the values are to be translated in terms of irrigation interval in days between two consecutive irrigations.

7.3 Plant indices approach

The plant in one form or the other expresses water deficits in the soil, since it is the one, which is affected by the water, stress. Any plant character, related directly or indirectly to water deficits and which responds readily to the integrated influences of soil water, plant and environmental parameters may serve as a criterion for timing irrigation to crops.

7.3.1 Visual plant symptoms

In this method the visual signs of plants are used as an index for scheduling irrigations. For instance, plant wilting, drooping, curling and rolling of leaves in maize is used as indicators for scheduling irrigation. Change in foliage colour and leaf angle is used to time irrigations in beans. Water stress in some crops leads to appearance of carotenoid (yellow and orange colour) and anthocyanin pigments; shortening of internodes in sugarcane and cotton; retardation of stem elongation in grapes; leaf abscission and lack of new growth and redness in terminal growth points of almond, which can be used as indices for scheduling irrigations to crops.

7.3.2 Soil-cum-sand mini-plot technique

This method is also referred to profile modification technique and is commonly used for scheduling irrigations to crops. The principle involved in this technique is to reduce artificially available water holding capacity of soil profile (i.e., effective root zone depth) in the mini-plot by mixing sand with it. When this is done, plants growing on the sand mixed plot show wilting symptoms earlier than in the rest of the field. An area of 1.0 x 1.0m is selected in the field and a pit of 1.0m depth is excavated. About 5% of sand by volume is added to the dug up soil and mixed well. The pit is then filled back with the mixture and while filling up, every 15 cm layer is well compacted, so that the soil in the pit retains the original bulk density as that of surrounding soil. Crop is sown normally and is allowed to grow as usual with the rest of the field. As and when the plants in the mini-plot show wilting symptoms it is taken as a warning of impending water need and cropped field is irrigated.

7.3.3 Plant population

Increase in plant population by 1.5 to 2.0 times that of optimum in some representative spots of (1 m × 1m area) in the cropped field alternative to mini-plot technique also serves as a reliable index for scheduling irrigation to crops. This happens because when more plants are there per unit area, the available water within that zone is depleted rapidly as compared to other area wherein optimum number of plants is maintained per unit area. This result in drooping or wilting of plants earlier, which can be taken as an indication of water deficits and accordingly irrigations are scheduled to crops.

7.3.4 Rate of growth

Growth of a plant is dependent on turgour, which in turn is dependent on a favourable soil water balance. So, fluctuations in the water balance are reflected by parallel fluctuations in the growth rate of expanding organs. Stem elongation is markedly reduced when the available soil moisture level approaches the critical level, but, accelerates again after irrigation. In cotton elongation of internodal length by 0.5 to 1.0cm day^{-1} indicated the approach of critical available soil moisture level. Likewise in oranges, when the growth rate of fruit circumference falls below 0.2 to 0.3mm d^{-1}, irrigations have to be scheduled.

7.3.5 Relative water content

This concept was proposed by Weatherly (1950). It is the actual water content of the leaf or plant when sampled relative to water at saturation or turgid. It is expressed as relative water content (RWC) and is calculated as follows:

$$RWC = \frac{\text{Fresh weight} - \text{Dry weight}}{\text{Turgid weight} - \text{Dry weight}}$$

It depends on the lag between evaporative demand of the atmosphere and the rate of water absorption by the roots and has been found to be one of the reliable indices of plant water stress for scheduling irrigation. For instance the critical RWC level (below which reduction in growth occurs) for cotton and sesame was 72% and 75% respectively. Whenever the plant exhibits these values, irrigations can be scheduled. However, the main drawback is it is tedious and time consuming.

7.3.6 Plant water potential

This method measures the energy status of plant water analogous to the tension of film water in the soil, serves as a better index of physiological and bio-chemical phenomena occurring in the plant. Plant or leaf water potential can be precisely measured either by a pressure bomb or pressure chamber apparatus *in-situ* or by the dye method in the laboratory. The critical plant water potential values for cotton below which yield reductions are expected were 1.2 to 1.25 MPa throughout the crop life. While for sunflower they were 1.0, 1.2 and 1.4 MPa at vegetative, pollination and seed formation, respectively.

7.3.7 Canopy temperature

Plant temperature or canopy temperature adequately reflects the internal water balance of the plant, and can be used as a potential indicator for scheduling irrigation to crops. It can be measured by several instruments, which are commercially available *viz.,* porometer, infrared thermometer etc. For maize, it is shown that if the canopy temperature rises to more than 0.7°C over ambient temperature during 1330 to 1400 hour irrigations need to be scheduled. However, tomato showed no sensitivity to water stress with respect to temperature variations.

7.3.8 Indicator plants

Use of indicator plants as a guide for scheduling irrigations. Crops like sunflower and tomato are highly sensitive to water stress which will show stress symptoms earlier than other stress tolerating crops. Hence to know the stress symptoms earlier, such sensitive crops are planted in random in the field and based on the stress symptoms noticed in such plants, scheduling of irrigation can be made. This approach is called as indicator plant technique in scheduling irrigation.In wheat, scheduling irrigations on the basis of wilting symptoms in maize and sunflower gave the highest grain yields.

7.3.9 Critical growth stagesapproach

For each crop, there are some growth stages at which moisture stress leads to irrevocable yield loss. These stages are known as critical period or moisture sensitive period.

Under limited water supply condition and to manage the water scarcity situation, irrigation is to be scheduled at moisture sensitive stages and irrigation can be skipped at non sensitive stages.

Moisture sensitive stages of some horticultural crops are given below.

Agricultural crops	Moisture sensitive period
Rice	Panicle initiation, Flowering
Wheat	Crown root initiation, Jointing, Milking
Sorghum	Seeding, Flowering
Maize	Silking, Tasseling
Pearl millet	Panicle initiation, Flowering
Finger Millet	Panicle initiation, Flowering
Groundnut	Rapid flowering. Pegging, Early pod formation
Sunflower	2 weeks before and after flowering
Sesamum	Blooming to maturity

Pulses	Flowering, Pod formation
Sugarcane	Formative stage
Cotton	Flowering, Boll development
Onion	Bulb formation to maturity
Tomato	Flowering and fruit setting
Chilies	Flowering
Cabbage	Head formation to maturity
Peas	Flowering and pod formation
Beans	Flowering and pod setting
Potato	Tuber initiation and maturity
Citrus	Flowering, fruit set and enlargement
Mango	Flowering
Coffee	Flowering and fruit development

QUESTIONS

I. Fill in the Blanks

1. The oldest and simple methods of determining the soil moisture content is ________________
2. ________________ used for irrigation scheduling based on soil moisture tension
3. Crops like _________ and ________ are highly sensitive to water stress and used as indicator crops.
4. Crop wilting, drooping, curling and rolling of leaves are ________________ used as indicators for scheduling irrigation.

II. Answer the fallowing questions

1. Define: Irrigation scheduling
2. Give details on soil cum sand mini plot techniques
3. List out plant criteria used for irrigation scheduling
4. Soil water regime approach for irrigation scheduling
5. What is the lysimeter and explain its function
6. What is IW/CPE ratio?
7. Plant indices approach for irrigation scheduling
8. Define: Relative Water Content

11

IRRIGATION STRUCTURES IN WETLANDS AND IRRIGATED DRYLANDS

1. IRRIGATION STRUCTURES

- Irrigation structures are useful to transfer water from source (dams, tanks, wells, etc.) to the required place.
- Good irrigation structures are an essential part of an irrigation layout. Efficient structures will save labour, land and water.

Types of irrigation structures

Conveyance structures	***Water control structures***	***Crossing structures***	***Erosion control structures***
1) Surface channels Unlined earthen channels Lined channels	1) Diversion boxes Two way Three way	1) Siphons 2) Culverts	1) Drop structures 2) Chute spill ways

2. CONVEYANCE STRUCTURES

- These are used to carry water from the source to the field (e.g. canals, channels, pipelines etc.).
- Canals are larger in size whereas channels are smaller in size.

2.1 Open channels (surface channels)

They are grouped into three types

1. Earthen channels (e.g.) canal, field channels and uncovered flumes
2. Lined channels
3. Pre-cast concrete cement channels

2.1.1 Earthen channels

- Earthern channel is the most commonly used water conveyance structure.
- The advantage is the low initial cost.
- Earthen channels should be built with stable side slopes and with banks strong to carry the required flow of water safely.
- They should carry the design discharge at non-erosive velocities.
- When earthen channels are to be used on deep slopes, it is necessary to control the gradients and thus the velocity by constructing drop structures or by building up the channel bed.
- Earthen channels require continuous maintenance to control weed growth.

2.1.2 Lined channels

- Earthen channels are lined with impervious materials to prevent excessive seepage and growth of weeds in channels.
- Lining of water courses will provide nearly equal distribution of water amongst all the farmers in the command area.
- Lining will reduce the labour cost in maintenance and almost eliminate water logging caused by seepage in water course.
- More land will be available for cultivation.
- Concrete and brick or stone masonry is commonly used for lining irrigation channels.

2.1.3 Pre-cast concrete channel sections

Could also be used for lining channel:

- These are concrete units manufactured at a central place and placed in the channel.
- After laying the sections their joints are filled with cement mortar.
- They are made in lengths of one meter or more and are generally 30 to 45 cm in diameter.

2.2 Underground pipe lines

- This system consists of buried pipes for conveying water to different points on the farm.
- Since the pipes are laid underground, cultivation can be done above the pipeline; no culverts or other structures are required.
- When properly installed they have long life and low maintenance costs.
- They are leak proof.
- Initial cost will be more.
- Under canal irrigation due to insufficient head available, open channels are usually preferred.

3. WATER CONTROL AND DIVERSION STRUCTURES

- These structures are necessary to control the irrigation water.
- Diversion boxes, turn outs, check gates, portable check dams, are some of the water control diversion structures.

3.1 Diversion boxes

- In carrying water to different farms, or to different parts of the same farm, the water course or main channel has to be diverted into proper channel.
- Two way or three way and four way diversions made of check gates at the channel junction will permit the operation of diverting all or any predetermined portion of the stream through the diversions.

3.2 Turnouts

- When water is to be taken from a lateral channel in to a field distribution channel or from a channel into a field, a turn out is used.
- It may be portable or built in. The most common turnouts are box turnouts, spills and siphon tubes.

3.3 Check gates

- Check gates are placed at intervals along the channel to keep a satisfactory water level for applying water to the field.
- They form an adjustable dam to control the elevation of the water surface upstream.

3.4 Portable check dams

- They are made up of canvas or plastic or sheet metal. Sheet metal dam provides greater durability than canvas or plastic dams.
- The dam should be driven about 7 to 10 cm deep into the bank to prevent seepage around the sides and bottom.
- To prevent rusting the sheet metal dams are always painted.

4. CROSSING STRUCTURES

- When irrigation channels are carried across roads, hillsides and natural depressions or drainage ways, the cross structures are essential.
- Siphons are used to carry irrigation channels across natural farm obstructions.
- When the siphon is laid over a drainage way or depression it is often called a flume.
- When it crosses a ridge, high way or bund it is called a culvert.

5. SOIL EROSION CONTROL STRUCTURES

- Drop structures and chute drops are used to prevent erosion in field channels.
- Severe erosion will occur in earth channels if structures to control the slope is not provided.

QUESTIONS

I. Fill in the blanks

1. Conveyance structures are used to carry water from the ______ to the _____
2. ________ channel is the most commonly used water conveyance structure.
3. Diversion boxes, turn outs, check gates are some of the ____________ structures.
4. Drop structures and chute drops are used to prevent __________in field channels.

II. Answer the following questions

1. Brief about irrigation structures
2. List out different types of irrigation structures with examples

12

TMC, CUSEC, CUMEC, IRRIGATION EFFICIENCIES AND CALCULATION

1. **TMC:**Thousand Million Cubic feet (1,000,000,000 = 10^9 = 1 Billion)
2. **CUSEC:** Cubic Feet per Second (28.317 litres per seconds)
3. **CUMEC:** Cubic Metres per second (1000 litres per seconds)

4. EFFICIENCIES

- An efficient irrigation system implies effective transfer of water from source to the field with minimum possible loss.
- The objective of the efficiency concept is to identify the nature of water loss and to decide the type of improvements in the system.
- Evaluation of performance in terms of efficiency is prerequisite for proper use of irrigation water.

4.1 Irrigation efficiency

It is defined as the ratio of water output to the water input, *i.e.* the ratio or percentage of the irrigation water consumed by the crop of an irrigated farm or field or project to the water delivered from the source.

$$E_i = \frac{W_c}{W_r} \times 100$$

Where,

E_i: Irrigation efficiency (%)

W_c: Irrigation water consumed by crop during growth period in an irrigation project

W_r: Water delivered from project during the growth period of crops

In most of the irrigation projects, the irrigation efficiency ranges between 12 to 34%.

4.2 Water conveyance efficiency

It is a measure of efficiency of water conveyance system from canal network to watercourses and field channels. It is the ratio of water delivered infields at the outlet head to that diverted into the canal system from the river or reservoir. Water losses occur in conveyance from the point of diversion till it reaches the farmer's fields can be evaluated by water conveyance efficiency, as under:

$$E_c = \frac{W_f}{W_d} \times 100$$

Where,

E_c: Water conveyance efficiency (%)

W_f: Water delivered to the farm by conveyance system (at field supply channel)

W_d: Water introduced into the conveyance system from the point of diversion

Water conveyance efficiency is generally low; about 21% losses occur in earthen water courses only.

4.3 Water application efficiency

It is a measure of efficiency of water application in the field. It is the ratio of volume of water that is stored in the root zone of crops and ultimately consumed by transpiration or evaporation or both to the volume of water actually delivered at the field.

Alternatively, it may be defined as the percentage of water applied that can be accounted for an increase in soil moisture in soils as occupied by the principal rooting system of the crop. It is also termed as farm efficiency as it takes into account water lost in application at the farm.

$$E_a = \frac{W_x}{W_f} \times 100$$

Where,

E_a: Water application efficiency (%)

W_s : Irrigation water stored in the root zone of farm soil

W_f: Irrigation water delivered to the farm (at field supply channel)

In general, water application efficiency decreases as the amount of water during each irrigation increases. Water losses due to inefficient application of water in the field vary from 28 to 50 %.

4.4 Water use efficiency

Having conveyed water to the point of use and having applied it, the next efficiency concept of concern is the efficiency of water use. It is expressed in kg/ha cm. The proportion of water delivered and beneficially used on the project can be calculated using the following formula

$$E_u = \frac{W_u}{W_d} \times 100$$

Where,

E_u : Water use efficiency (%)

W_u : Water beneficially used

W_d : Water delivered

Water use efficiency is also defined as (i) crop water use efficiency and (ii) field water efficiency.

4.4.1 Crop water use efficiency

It is the ratio of yield of crop (*Y*) to the amount of water depleted by crop in evapotranspiration (*ET*).

$$\text{CWUE} = \frac{Y}{ET}$$

Where,

CWUE : Crop water use efficiency

Y : Crop yield

ET : Evapotranspiration

CWUE is otherwise called consumptive water use efficiency. It is the ratio of crop yield (*Y*) to the sum of the amount of water taken up and used for crop growth (*G*), evaporated directly from the soil surface (*E*) and transpired through foliage (*T*) or consumptive use (Cu)

$$\text{CWUE} = \frac{Y}{G+E+T}$$

Where,

$$(G + E + T) = \text{Cu}$$

In other words *ET* is Cu since water used for crop growth is negligible.

$$\text{CWUE} = \frac{Y}{CU}$$

It is expressed in kg/ha/mm or kg/ha/cm.

4.4.2 Field water use efficiency

It is the ratio of yield of crop (*Y*) to the total amount of water used in the field.

$$\text{FWUE} = \frac{Y}{WR}$$

Where,

FWUE: Field water use efficiency

WR: Water requirement

This is the ratio of crop yield to the amount of water used in the field (*WR*) including growth (*G*), direct evaporation from the soil surface (*E*), transpiration (*T*) and deep percolation loss (*D*).

$$\text{FWUE} = \frac{Y}{G+E+T+D}$$

$$G + E + T + D = WR$$

It is expressed in kg/ha/mm (or) kg/ha/cm

Of the two indices defined, the crop water use efficiency is more of research value and the field water use efficiency has grater practical importance for planners and farmers.

4.5 Water storage efficiency

It is defined as the ratio of the water stored in the root depth by irrigation to the water needed in the root depth to bring it to the field capacity. Also termed as water storage factor.

$$E_s = \frac{W_s}{W_w} \times 100$$

Where,

E_s : Water storage efficiency (%)

W_s : Water stored in the root zone during the irrigation

W_w : Water needed in the root zone prior to irrigation, i.e.,field capacity available moisture.

4.6 Water distribution efficiency

Expression for distribution efficiency to evaluate the extent to which the water is uniformly distributed is as follows:

$$E_d = \left(1 - \frac{d}{D}\right) 100$$

Where,

E_d : Water distribution efficiency (%)

D : Average numerical deviation in depth of water stored from average depth stored during irrigation

D : Average depth of water stored along the run during irrigation

A water distribution efficiency of 80% means that 10% of water was applied in excess and consequently 10% was deficient in comparison to the average depth of application.

4.7 Consumptive use efficiency

It is defined as the ratio of consumptive water use by the crop of irrigated farm or project and the irrigation water stored in the root zone of the soil on the farm or the project area. After irrigation water is stored in the soil, it may not be available for use by the crop because water may evaporate from the ground surface or continuously move downward beyond the root zone as it may happen in a wide furrow spacing. The loss of water by deep penetration and by surface evaporation following an irrigation is evaluated from the following expression:

$$\mathrm{E_{cu}} = \frac{W_{cu}}{W_d} \times 100$$

Where,

E_{cu} : Consumptive use efficiency (%)

W_{cu} : Normal consumptive use of water

W_d : Net amount of water depleted from root zone soil

Consumptive use efficiency is useful in explaining the difference in crop response from different methods of irrigation.

5. CALCULATION OF IRRIGATION WATER

The major aspects to be considered in irrigation management are

- How much to irrigate?
- When to irrigate?

5.1 How much to irrigate?

The amount of water to be applied at each irrigation depends on

- Root zone depth
- Soil type
- Water deficit before irrigation
- The amount of rainfall
- Salt content in the soil

When no rainfall is likely to be received, the net quantity of water to be applied will be equal to the moisture deficit in the soil.

5.1.1. Moisture deficit

$$d = \sum_{i=1}^{n} \frac{Pfci - Pwi}{100} \times \text{Di} \times \text{Asi}$$

Where,

d : Total moisture deficit in the root zone depth, D

Pfci : Field capacity of ith layer on oven dry weight basis

Pwi : Actual moisture content of ith layer on oven dry weight basis

Di : Depth of ith layer

Asi : Apparent specific gravity of ith layer

n : Total number of layers in the root zone

5.1.2 Available moisture storage capacity of the soil

$$\text{Storage capacity} = \frac{FC - WP}{100} \times AS \times D$$

Where,

FC: Field capacity of the soil on oven dry weight basis

WP: Wilting point of the soil on oven dry weight basis

AS: Apparent specific gravity of the soil

D: Depth of root zone

5.2 When to irrigate? (Irrigation interval)

$$\text{Irrigation interval (days)} = \frac{\text{Depth of irrigation (mm)}}{\text{Daily consumption use (mm)}}$$

$$\text{Depth of irrigation (m)} = \frac{BD \times D\,\text{(m)}}{\text{Unit weight of water}} \times (FC - AM)$$

Where,

BD: Bulk density (g/cc)

D: Root zone depth (m)

FC: Field capacity (expressed as ratio)

AM: Available moisture (expressed as ratio)

or

$$= \frac{\text{Volume of water (m}^3\text{)}}{\text{Area of land (m}^2\text{)}}$$

5.3 Time required to irrigating an area

The time required to irrigate in an area depends on,

- The magnitude of discharge
- Quantity of water to be applied
- Irrigation efficiency and area to be irrigated

It can be calculated by using the formula

$$Iqt = Ad$$

Where,

I : Irrigation efficiency

q : Discharge (Cusec (or) Cumec/hr)

t : Time (hrs)

A : Area (m^2)

d : Moisture deficit in soil

5.4 Area covered

$$\text{Area irrigated} = \frac{\text{Volume of water available (m}^3\text{)}}{\text{Volume of water required/ha (m}^3\text{)}}$$

- Required total discharge

QUESTIONS

I. Expand

1. TMC
2. CUSEC
3. CUMEC

II. Fill in the blanks

1. Irrigation efficiency of irrigation projects, ranges between __________%.
2. Water losses from the earthen water courses about _________ %.
3. Unit for crop water use efficiency is __________.
4. ____________ is more important irrigation indices for planners and farmers.

III. Answer the following questions

1. Irrigation efficiency
2. Water conveyance efficiency
3. Water application efficiency
4. Water use efficiency
5. Crop water use efficiency
6. Field water use efficiency
7. Water storage efficiency
8. Water distribution efficiency
9. Consumptive use efficiency.

13

WEEDS - DEFINITION, CLASSIFICATION AND CHARACTERISTICS, HARMFUL AND BENEFICIAL EFFECTS OF WEEDS

1. DEFINITIONS

- Weed is a plant growing out of place.
- Weeds are unwanted and undeserved plants that interfere with the utilization of the land and water resources and thus adversely affect crop production and human welfare.
- Plant out of its place or a plant growing where it is not desired is a weed. This definition was given by Buchholtz in 1967. Example: Cumbu is a weed in pulse, pulse is a weed in cumbu, tomato is a weed in brinjal field.

They are unwanted not useful, persistent and prolific, effectively competing with the beneficial and desirable crop plants for space, nutrients, sunlight, and water, interfere with agricultural operations and thereby reducing the yield and quality of produce.

Weeds are all types of undesirable plants i.e. sedges grass, broad leaved weeds, aquatic plants, trees and parasitic flowering plants (*Striga* / *Orobanche*) affecting crop area and non- crop area (industrial side, road side, railway line, water tank, irrigation channel etc.).

Weeds are unwanted, pernicious and harmful plants which interfere with agriculture operations, increase labour, add to the cost of cultivation and reduce yield of crops. Weed has no species. but name was suggested as a useless and harmful plant that persistently grows where it is quite unwanted.

Weed is defined as the unwanted, undesirable plant growing out of their proper place which interfere with the utilization of natural resources, prolific, persistent, competitive, harmful and even poisonous in nature and can grow in adverse climatic conditions. (Jethro Tull: father of weed science)

Weediness: Weediness is defined as the state or condition of a field, flower beds, lawns and so forth in which there is an abundance of weeds.

2. CLASSIFICATION OF WEEDS

Weeds can be grouped for the convenience of planning, interpreting and recording control measures against them. Weeds belonging to any group of these classes have specific mode of propagation, dispersal and persistence. Out of 2,50,000 plant species, weeds constitute about 250 species, which are prominent in agricultural and non-agricultural system. Under world conditions about 30000 species is grouped as weeds. Out of these 18,000 species can causes damage to the crops.

2.1 Based on life span

Based on life span (Ontogeny), weeds are classified as Annual weeds, Biennial weeds and Perennial weeds.

2.1.1 Annual weeds

Those that live only for a season or year and complete their life cycle in that season or year is called annuals.

These are small herbs with shallow roots and weak stem. Produces seeds in profusion and the mode of propagation is commonly through seeds. After seeding the annuals die away and the seeds germinate and start the next generation in the next season or year following.

Mostly common field weeds are annuals. The examples are

(a) **Monsoon annual:** *Commelina benghalensis, Boerhaavia erecta*

(b) **Winter annual:** *Chenopodium album*

2.1.2 Biennials

It completes the vegetative growth in the first season, flower and set seeds in the succeeding season and then die. These are found mainly in non-cropped areas.

Example: *Alternanthera echinata, Daucus carota*

2.1.3 Perennials

Perennials live for more than two years and may live almost indefinitely. They adapted to withstand adverse conditions. They propagate not only through seeds but also by underground stem, root, rhizomes, tubers etc. And hence they are further classified into

Simple perennials: Plants propagated only by seeds. Example: *Sonchus arvensis*

Bulbous perennials: Plants which possess a modified stem with scales and reproduce mainly from bulbs and seeds. Example: *Allium* sp.

Corm perennials: Plants that possess a modified shoot and fleshy stem and reproduce through corm and seeds. Example: *Timothy* sp.

Creeping perennials: Reproduced through seeds as well as with one of the following.

(a) **Rhizome:** Plants having underground stem – *Sorghum halapense*

(b) **Stolon:** Plants having horizontal creeping stem above the ground – *Cynodon dactylon*

(c) **Roots:** Plants having enlarged root system with numerous buds – *Convolvulus arvensis*

(d) **Tubers:** Plants having modified rhizomes adapted for storage of food – *Cyperus rotundus*

2.2 Based on ecological affinities

2.2.1 Wetland weeds

They are tender annuals with semi-aquatic habit. They can thrive as well under waterlogged and in partially dry condition. Propagation is chiefly by seed.

Example: *Ammania baccifera, Eclipta alba*

2.2.2 Garden land weeds (Irrigated lands)

These weeds neither require large quantities of water like wetland weeds nor can they successfully withstand extreme drought as dryland weeds

Example: *Trianthema portulacastrum, Digera arvensis*

2.2.3 Dry lands weeds

These are usually hardy plants with deep root system. They are adapted to withstand drought on account of mucilaginous nature of the stem and hairiness.

Example: *Tribulus terrestris, Convolvulus arvensis*

2.3 Based on soil type (Edaphic)

2.3.1. Weeds of black cotton soil

These are often closely allied to those that grow in dry condition. Example: *Aristolochia bracteata*

2.3.2 Weeds of red soils

They are like the weeds of garden lands consisting of various classes of plants. Example: *Commelina benghalensis*

2.3.3 Weeds of light, sandy or loamy soils

Weeds that occur in soils having good drainage. Example: *Leucas aspera*

2.3.4 Weeds of laterite soils

Example: *Lantana camara, Spergula arvensis*

2.4. Based on place of occurrence

2.4.1. Weeds of crop lands

The majority of weeds infest the cultivated lands and cause hindrance to the farmers for successful crop production. Example: *Philaris minor* in wheat

2.4.2 Weeds of pasture lands

Weeds found in pasture / grazing grounds. Example:*Indigofera enneaphylla*

2.4.3 Weeds of waste places

Corners of fields, margins of channels etc., where weeds grow in profusion. Example: *Gynandropsis pentaphylla*, *Calotropis gigantea*

2.4.4 Weeds of playgrounds, road-sides

They are usually hardy, prostrate perennials, capable of withstanding any amount of trampling. Example:*Alternanthera echinata, Tribulus terestris*

2.5 Based on Origin

2.5.1 Indigenous weeds

All the native weeds of the country are coming under this group and most of the weeds are indigenous. Example:*Acalypha indica, Abutilon indicum*

2.5.2 Introduced or Exotic weeds

These are the weeds introduced from other countries. These weeds are normally troublesome and control becomes difficult. Example: *Parthenium hysterophorus, Phalaris minor, Acanthospermum hispidum*

2.6 Based on cotyledon number

Based on number of cotyledons it possess it can be classified as dicots and monocots.

2.6.1 Monocots

Example: *Panicum flavidum, Echinochloa colona*

2.6.2 Dicots

Example: *Crotalaria verucosa, Indigofera viscosa*

2.7 Based on soil pH

Based on pH of the soil the weeds can be classified into three categories.

2.7.1 Acidophile

Acid soil weeds Example: *Rumex acetosella*

2.7.2 Basophile

Saline & alkaline soil weeds Example:*Taraxacum stricta*

2.7.3. Neutrophile

Weeds of neutral soils Example: *Acalypha indica*

2.8 Based on morphology

Based on the morphology of the plant, the weeds are also classified in to three categories. This is the most widely used classification by the weed scientists.

2.8.1. Grasses

All the weeds come under the family Poaceae are called as grasses which are characteristically having long narrow spiny leaves.

Example:*Echinocloa colonum, Cynodon dactylon*

2.8.2. Sedges

The weeds belonging to the family Cyperaceae come under this group. The leaves are mostly from the base having modified stem with or without tubers. Example: *Cyperus rotundus, Fimbrystylis miliaceae*

2.8.3 Broad leaved weeds

This is the major group of weeds as all other family weeds come under this except that is discussed earlier. All dicotyledon weeds are broad leaved weeds. Example: *Flavaria australacica, Digera arvensis.*

Phots of major grasses, sedges and broad leaf weeds are enclosed in Annexure I.

2.9 Based on nature of stem

Based on development of bark tissues on their stems and branches, weeds are classified as woody, semi-woody and herbaceous species.

2.9.1. Woody weeds

Weeds include shrubs and undershrubs and are collectively called brush weeds.
Example: *Lantana camera, Prosopis juliflora*

2.9.2 Semi-woody weeds

Example: *Croton sparsiflorus*

2.9.3 Herbaceous weeds

Weeds have green, succulent stems are of most common occurrence around us.
Example: *Amaranthus viridis*

2.10 Based on specificity

Besides the various classes of weeds, a few others deserve special attention due to their specificity. They are;

(a) Poisonous weeds

(b) Parasitic weeds

(c) Aquatic weeds

2.10.1 Poisonous weeds

The poisonous weeds cause ailment on livestock resulting in death and cause great loss. These weeds are harvested along with fodder or grass and fed to cattle or while grazing the cattle consumes these poisonous plants. Example: *Datura fastuosa, D. stramonium* and *D. metel* are poisonous to animals and human beings. The berries of *Withania somnifera* and seeds of *Abrus precatorius* are poisonous.

2.10.2 Parasitic weeds

The parasite weeds are either total or partial which means, the weeds that depend completely on the host plant are termed as total parasites while the weeds that partially depend on host plant for minerals and capable of preparing its food from the green leaves are called as partial parasites.

Those parasites which attack roots are termed as root parasites and those which attack shoot of other plants are called as stem parasites. The typical examples of different parasitic weeds are

Total root parasite - *Orabanche cernua* on Tobacco

Partial root parasite - *Striga lutea* on sugarcane and sorghum

Total stem parasite - *Cuscuta chinensis* on leucerne and onion

Partial stem parasite - *Cassytha filiformis* on orange trees and *Loranthus longiflorus* on mango and other trees.

2.10.3 Aquatic weeds

Unwanted plants, which grow in water and complete at least a part of their life cycle in water are called as aquatic weeds. They are further grouped into four categories as submersed, immersed, marginal and floating weeds.

2.10.3.1 Submersed weeds

These weeds are mostly vascular plants that produce all or most of their vegetative growth beneath the water surface, having true roots, stems and leaves.

Example: *Utricularia stellaris, Ceratophyllum demersum.*

2.10.3.2 Immersed weeds

These plants are rooted in the bottom mud, with aerial stems and leaves at or above the water surface. The leaves are broad in many plants and sometimes like grasses. These leaves do not rise and fall with water level as in the case of floating weeds. Example: *Nelumbium speciosum, Jussieua repens*

2.10.3.3 Marginal weeds

Most of these plants are immersed weeds that can grow in moist shoreline areas with a depth of 60 to 90 cm water. These weeds vary in size, shape and habitat. The important genera that come under this group are; *Typha, Polygonum, Cephalanthus, Scirpus, etc.*

2.10.3.4. Floating weeds

These weeds have leaves that float on the water surface either singly or in cluster. Some weeds are free floating and some rooted at the mud bottom and the leaves rise and fall as the water level increases or decreases. Example: *Eichhornia crassipes, Pistia stratiotes, Salvinia, Nymphaea pubescens*.

3. CHARACTERISTICS OF WEEDS

- Weeds have rapid seedling growth and ability to reproduce when young e.g. Redroot Pigweed can flower and reproduce when it is less than eight inches tall.
- Weeds have quick maturation period or takes only a short time in the vegetative phase e.g. *Canadathistle* can produce mature seeds in two weeks after flowering. Example: Russianthistle seeds can germinate very quickly between 28^0-110 ^{0}F in late spring.
- Weeds may have dual mode of reproduction. Most weeds are angiosperms & reproduce by seeds. Many reproduce vegetatively too.
- Weeds have environmental plasticity. Many weeds are capable of tolerating and growing under a wide range of climatic and edaphic conditions.
- Weeds are often self-compatible but self-pollination is not obligatory.
- If a weed is cross pollinated, this is accomplished by non- specified flower visitors or by wind.
- Weeds resist detrimental environmental factors (Most crop seed rot, if they do not germinate shortly after planting) and weed seeds resist decay for long periods in soil and remain dormant for longer duration.
- Weed seeds exhibit several kinds of dormancy and escape the rigors of environment and germinate when conditions are more favorable for their survival. Many weeds have no special environment requirements for germination.
- Weeds often produce seeds of same size and shape as the crop seed, making physical separation difficult and facilitating spread by men.
- Some annual weeds produce more than one seed crop per year and seed is produced as long as growing conditions permit.

- Each generation is capable of producing large number of seeds per plant and seed is produced over a wide range of environmental conditions.
- Many weeds have specially adapted long and short range seed dispersal mechanisms.
- Roots of some weeds are able to penetrate and emerge from deep in the soil, while most roots are in the upper foot of the soil; *Canada thistle* roots routinely penetrate 3-6 feet and fieldbindweed roots upto 10 feet deep. Roots and rhizomes are capable of growing many feet per year.
- Roots and other vegetative parts of perennial weeds are vigorous with large food reserves, enabling them to withstand environmental stress.
- Perennials have bitterness in the lower stem nodes or in rhizomes and roots and, if severed vegetative organs will quickly regenerate in to whole plant.
- Many weeds have adaptation mechanism that repels grazing, such as spines, odd taste or odour.
- Weeds have great competitive ability for nutrients, light and water and can compete by special means, (e.g. Rosette formation, climbing growth and allelopathy.)
- Weeds are ubiquitous (present everywhere). They exist everywhere, where we practice agriculture.
- Weeds resist control, including herbicides.

4. HARMFUL EFFECTS OF WEEDS

Of the total annual loss of agricultural produce from various pests in India, weeds account for 45%, insects 30%, diseases 20% & others 5%.

The losses due to weeds depends on

1. Type of weed
2. Severity of Infestation
3. Duration of infestation
4. Competitive ability of the crop plants with weeds
5. Climatic conditions which effect the growth of the crop and the weed

- Weeds have serious impacts on agricultural production. It is estimated that in general weeds cause 5% loss in agricultural production in most of developed countries, 10% loss in less developed countries and 25% loss in least developed countries.
- In India, yield losses due to weeds are more than those from pest and diseases. Yield losses due to weeds vary with the crops. Every crop is exposed to severe competition from weeds. Most of these weeds are self-sown and they provide competition caused by their faster rate of growth in the initial stages of crop growth. In some crops, the yields are reduced by more than 50% due to weed infestation. These loses caused by weeds in some of the important crops are given in the following table.

- Weeds compete with crops for water soil, nutrients, light, and space, and thus reduce the crop yields. An estimate shows that weeds can deprive the crops 47% N, 42% P, 50% K, 39% Ca and 24% Mg of their nutrient uptake.
- Weeds are also act as alternate hosts that harbor insects, pests and diseases and other micro-organisms. Alternate hosts of some of the pest and diseases

Crop	Pest	Alternate host
Sweet potato	Sweet potato weevil	*Convolvulus arvensis*
Potato	Stalk borer, beetles and cutworms	*Chenopodium album*

Some weeds release into the soil inhibitors of poisonous substances that may be harmful to the crop plants, human beings and livestock. Health problems caused by weeds to humans,

Health problem	Weed
Hay fever and Asthma	Pollen of *Ambrosia* and *Franseria*
Dermatitis	*Parthenium, Ambrosia*
Itching and Inflammation	*Utrica* sp.
African sleeping sickness	Brush weeds
Malaria, encephalitis and filaria caused by mosquito	Aquatic weeds like *Pistialanceolate, Salviniaauriculata*

Weeds reduce the quality of marketable agricultural produce. Contamination of weed seeds of *Datura, Argemone, Brassica* etc., is harmful to human health and weed seeds present in the produce cause odd odour sometimes.

- Weeds not only reduce yield but also interfere with agricultural operations. Weeds make mechanical sowing a difficult process and render harvesting difficult, leading to increased expenditure on labour, equipment and chemicals for their removal.
- In aquatic environment, weeds block the flow of water in canals, water-transport system and drainage system, rendering navigation difficult. The dense growth of aquatic weeds pollutes water by deoxygenating it and killing the fished.
- Weeds are also a nuisance and a fire hazard along railway lines, roads, right-of-ways, airports, forest and industrial sites.

The following are the harmful effects of weeds on various activities of human beings, crops, livestock and related aspects.

1. Weed menace in agriculture
2. Weed menace in animal husbandry
3. Weed menace to human health
4. Weed menace to aquatic ecosystems
5. Weed menace to forest and pasture land
6. Weeds as alternate hosts to crop pests and diseases
7. Weeds reduce land value.
8. Weeds limits choice of crops

Weed menace in agriculture

Reduction of crop yield has a direct correlation with weed competition. Weeds compete for water, light, nutrients and space. Weeds compete for water in dry land and for nutrients in irrigated crops. It includes reduction in crop yields and production efficiency and erosion of crop quality.

Reduction in crop yields and production efficiency is direct effect due to weeds. It varies from 34.3% to 89.8% depending upon the crop. In rice (30-35%), wheat (15 -30%), Maize, sorghum, pulses, oilseeds (18-85%), sugarcane (38.8%), cotton (47.5%), sugar beet (48.4%) and onion (90.7%).

Beside the direct reduction in crop yields there are many indirect ways by which the weeds may be troublesome in agriculture. For example in weedy fields management practices become cumbersome. Harvest may be difficult when the field is invading with wild safflower (*Carthamus oxycantha*), Canada thistle (*Cirsium arvense*), cocklebur (*Xanthium strumarium*). Cowage (*Mucuna pruriens*) cause itching to the labour. Harvesting becomes troublesome when the field bindweed (*Convolvulus arvensis*) and morning glory (*Ipomoea* sp.) bind the crop plants together. The weeds at harvest may increase the excessive wear and tear of the farm machines and there by increased the cost of production to separate the weed seeds from the grain and other farm produce

Erosion of crop quality

Weeds may reduce the quality of farm produce in many ways. In dry land agriculture weeds cause severe moisture stress and force the food grains to shrivel. The vegetables and fruits are discoloured and deshaped in the presence of weeds. Contamination of food grains with poisonous weed seeds fetches low price. Foundation or certified seed is rejected if weed seeds exceed 2% and also the market value is reduced. The quality of the sugarcane crop is also reduced due to the presence of the parasitic weed *Striga litura.* Whereas leaves of *Loranthus* (*Dendrophthoe falcate*) impair the quality of tea. Leaves of *Mikania micrantha* create problem in tea plantation. In cotton the dry weed fragments adhere to the lint and hinder its spinning process. In India Cocklebur (*Xanthium srtumarium*) reduce the quality of wool in sheep. Oil quality of mint was impaired by *Cirsium arvense*

Weed menace in animal husbandry

Milk gives odd smell when animal fodder crops are mixed with wild onion and wild garlic, *Cichorium intybus, Argemone mexicana.* Certain weeds cause sickness and death of animals due to high levels of alkaloids, tannins, oxalates, glucosites ornitrates. Death of herds of sheep occurs due its high oxalate content of *Halogeton glomeratus*. It is dominant weed of arid and semi arid region. Leaves of *lantana* cause acute photosensitivity and jaundice in animals was due to the toxic principle of "Lantradene". Puncture wine (*Tribulus terrestris*) a weed of dryland induces extra sensitivity to light in sheep and puncture of the animal skin. In Kashmir *rhododendron* cause diarrhea and showed blood strains in milk. *Crotalaria spp.* is fatal to chick where as sweet clover (*Melilotus alba*) contains "dicumarin" which act as anti blood coagulant.

Weeds as alternate hosts to crop pests and diseases

In rice crop *Echinochloa and Panicum* act as alternate host for stem borer, whereas in maize *Chenapodium album* acts a host for stalk borer and beetles and cut worms in case of tomato. *Crotalaria spp.* act as a main host for hairy caterpillar which effect castor crop.

Diseases

Wild mustard act as alternate host for club root / black leg along with cabbage, cauliflower and for stem rust wild oats as alternate host which is major disease of wheat, oat, barley.

Reduction in land value

Land unsuitable for economic crop production when land infested with *Cyperus rotundus*, *Cynadon dactylon*. Thousands of hectare of cultivable area in rice growing regions of India have been abandoned or not being regularly cultivated due to severe infestation of nutgrass (*Cyperus rotundus*) and other perennial grasses.

Weed menace to human health

Hay fever and asthma aggravated by pollens of *Ambrosia artemissifolia*, *Parthenium hysterophous*, Poison ivy (*Rhus* sp.). Common rag weed (*Ambrosia artemissifolia*) are responsible for respiratory problems and skin allergies (dermatitis). Aquatic weeds like water lettuce (*Pistia lanceolata*), salvinia (*Salivina auriculata*), alligator weed (*Alternanthera* sp.) act as alternate host and vectors of malaria, yellow fever, encephalitis, dengue fever and filariasis. Wheat flour contaminated with seeds of cocklebur gives bitter taste to the bread and irritates the gastric tract of the consumer. When the Mexican poppy seeds (*Argemene mexicana*) crushed with mustard seeds cause death and blindness of people. Milk from animals feed on the Mexican poppy weed can cause „Glaucoma" in humans. The *Argemone* toxicity is due to an alkaloid called "Sanguinarine" and 11-oxotriacontanoic acid.

Weed menace to aquatic ecosystems

Aquatic weeds make the appearance of water bodies repulsive and decline their recreational value. Weeds hinder the navigation, fishing and slow down the flow in irrigation channels Aquatic weeds upon decomposition emit offensive odors and pollute the drinking water bodies.

Weed menace to forest and pasture lands

In Karnataka and Andhra Pradesh large forest area come under the grip of *Parthenium hysterophorus*. In West Bengal 11% of high forest area and 38% of plantation forest have been affected by *Mikania micrantha* (mile-a-minute weed).

Limits choice of crops

Crops differ in their ability to compete with weeds. In many instances, the presence of a particular weed in the field limits the choice of crops to be grown. Heavy weed infestation renders some economically important crops, particularlypulses, Vegetables, Cotton, Jute and forage crops unsuitable or less suitable for cultivation.

5. BENEFICIAL EFFECTS

- In spite of all the difficulties caused by weeds, they can offer some beneficial properties, particularly when occurring at low densities. These aspects should be utilised in the farming system, although this may make organic management more complicated than chemical based systems. Some of the potential benefits of weeds are listed below:
- Helping to conserve soil moisture and prevent erosion. A ground cover of weeds will reduce the amount of bare soil exposed helping to conserve nutrients, particularly nitrogen which could otherwise be leached away, especially on light soils.
- Food and shelter can be provided for natural enemies of pests and even alternative food sources for crop pests. The actual presence of weed cover may be a factor in increasing effectiveness of biological control of pests and reducing pest damage.
- Weeds can also be valuable indicators of growing conditions in a field, for example of water levels, compaction and pH.
- Weeds can be an important source of food for wildlife, especially birds. Bird populations have been declining on farmland over the last few decades and leaving weeds as a resource has been shown to help revive bird populations.

Several weeds have been put to certain economic uses since ages. *Typha* and *Saccharum* sp. used for making ropes and thatch boards. *Cichorium intybus* roots are used for adding flavor to coffee powder. *Amaranthus viridis*, *Chenopodium album* and *Portulaca sp.* used as leafy vegetable. In north India, *Saccharum spontaneum* used in breeding programme for developing the noble canes.

Incorporation of *Crotalaria*, *Parthenium*, *Calotropis* and *Eichornia* reduced root knot nematode population in the soil as they exhibited nematicidal properties. Hariyali grass (*Cynodon dactylon*) and *Cenchrus ciliaris*, *Dichanthium annulatum, Eclipta* alba *weeds* of grass land serve as food for animals. People in China and Japan consume *Chlorella pyrenoides* (algae) as protein supplement. Weeds act as alternate host for predators and parasites of insect pests which feed on the weeds. For example *Trichogramma chilonis* feed upon eggs of caster semi looper which damage the castor plants.

Some weeds useful to identify the metals (Indicator geobotany) through satellite imageries. Eg: *Commelina sp.* (Copper), *Eichornia crassipes* (Copper, zinc, lead and cadmium in water bodies). Several species of weeds, *Tephorsia purpurea* and *Crotons parsiflora* in south India used as green manures. Whereas *Eichornia crassipes* and *Pistia stratiotes* are used for composting. *Argemone mexicana* used for reclamation of alkali soils. NO_2 and SO_2 air pollution determined by wild mustard and chick weed respectively.

Aquatic weeds are useful in paper, pulp and fiber industry. *Chenopodium album* used as mulch to reduce evaporation losses, where as *Agropyron repens* (quack grass) used to control soil erosion because of its prolific root system. Weeds like *Lantana camera*, *Amaranthus viridis*, *Chenopodium album*, *Eichhornia crassipes* used for beautification. *Agropyron repens* used for soil conservation, where as *Dicanthium annulatum* stabilizing field bunds. *Opuntia dellini* used as biological fence.

Some weeds have medicinal properties and used to cure snake bite (*Leucas aspera*), gastric troubles (*Calotropis procera*), skin disorders (*Argemone mexicana*) and jaundice (*Phyllanthus niruri*) and *Striga orobanchioides* to control diabetes. In addition to the above agarbathis (*Cyperus rotundus*), aromatic oils (*Andropogan sp.* & *Simbopogon sp.*) are prepared from weeds.

QUESTIONS

I. Fill in the blanks

1. Father of weed science ________
2. Weeds live only one a season or year is called as ________
3. Weeds complete their life cycle in water is called as ________
4. ________ weed has tender annuals with semi-aquatic habit
5. Weed growing in acid soil is called as ________

II. Choose the best answers

1. Significant examples for Acidophile
   (a) *Acalypha indica* (b) *Rumex acetosella*
   (c) *Indigofera viscosa* (d) None of the above
2. Total root parasite
   (a) *Orabanche cernua* (b) *Striga lutea*
   (c) *Cuscuta chinensis* (d) *Loranthus longiflorus*
3. Host plant of Cuscuta
   (a) Lucerne (b) Mango (c) Sorghum (d) Tobacco
4. Hay fever and Asthma caused by
   (a) *Parthenium* (b) *Salvinia auriculata*
   (c) Ambrosia (d) *Utrica sp.*

III. Answer the following questions

1. Define weed
2. Write harmful and beneficial effect of weeds
3. List out the weeds based on life spanand morphology
4. List out weeds based on soil type and specificity

14

CLASSIFICATION AND CHARACTERISTICS OF WEEDS OF DIFFERENT AGRO ECOSYSTEMS: LOWLAND, IRRIGATED UPLAND, RAINFED LAND AND WASTE LAND

1. CLASSIFICATION BASED ON ECOLOGY AFFINITY

- Wet land weeds
- Garden land weeds
- Dry land weeds

2. GENERAL CHARACTERS / BIOLOGY OF WETLAND WEEDS

- Adapted to water logged condition
- Have lower requirement for light and oxygen than dryland weeds
- Anaerobic condition favours most of the wetland weeds
- Mostly annual in nature
- Most of the weeds in wetland are heavy seed producer
- Ability to float in water is achieved by hollowness of the stem, spongy tissues and air pockets at base of the stem.

List of predominant and important weeds of wetland ecosystem

Sl. No.	Scientific Name	Family	Common Name	Tamil Name
	A. Sedges			
1.	*Cyperus difformis*	Cyperaceae	Small flower umbrella	Vattakorai
2.	*Cyperus iria*	Cyperaceae	Umbrella sedge	Oosikorai
3.	*Cyperus rotundus*	Cyperaceae	Purple nut sedge	Koraikhizhangu
4.	*Fimbristylis miliaceae*	Cyperaceae	Hoorah grass	Poongkorai
	B. Grasses			
1.	*Brachiaria mutica*	Gramineae/ Poaceae	Para grass	Neerpul
2.	*Echinochloa colona*	Gramineae/ Poaceae	Jungle rice	Kudiraivalipul

3.	*Echinochloa crus-galli*	Gramineae/ Poaceae	Barnyard grass	Varsnampul
4.	*Imperata cylindrica*	Gramineae/ Poaceae	Cogon grass	—
5.	*Panicum repens*	Gramineae/ Poaceae	Torpedo grass	Injipul
	C. Broad leaved weeds			
1.	*Ammania baccifera*	Lythraceae	Blistering ammania	Neermel neruppu
2.	*Asteracantha longifolia*	Acanthaceae	Creeping spider ling	Neer mulli
3.	*Centella asiatica*	Umbelliferae	Indian penny wort	Vallarai
4.	*Eclipta alba*	Compositae	False daisy	Karisalanganni
5.	*Lippia nodiflora*	Verbanaceae	—	Poduthalai
6.	*Ludwigia parviflora*	Onagraceae	Creeping water primrose	Neer kirambu
7.	*Marsilea quadrifoliata*	Marsileaceae	The water fern	Arakeerai
8.	*Mimosa pudica*	Mimosaceae	The sensitive plant	Thotta sinungii
9.	*Monieria cuneifolia*	Scrophulariaceae	—	Neer pirami
10.	*Monochoria vaginalis*	Pontederiaceae	Thyme leaved grativola	Neer thamarai
11.	*Phyllanthus niruri*	Euphorbiaceae	Gripe weed	Keezha nelli
12.	*Rotala densiflora*	Lythraceae	Rotala	—
13.	*Sphaeranthus indicus*	Compositae	—	Kottai karanthai

3. GENERAL CHARACTERS/ BIOLOGY OF GARDENLAND WEEDS

The general morphological adaptation of gardenland weeds are:

- Weeds neither require moisture nor dry condition for survival
- Mostly biennials and some are perennial

List of predominant and important weeds of gardenland ecosystem

Sl. No.	Scientific Name	Family	Common Name	Tamil Name
	A. Sedges			
1.	*Cyperus bulbosus*	Cyperaceae	Bulb grass	Silandi
2.	*Cyperus rotundus*	Cyperaceae	Purple nut sedge	Korai kizhangu
	B. Grasses			
1.	*Chloris barbata*	Poaceae	Finger grass	Mayilkondaipul
2.	*Cynodon dactylon*	Poaceae	Bermuda grass	Arugampul
3.	*Dactyloctenium aegyptium*	Poaceae	Crow foot grass	Kakka kalpul

Sl. No.	Scientific Name	Family	Common Name	Tamil Name
4.	*Echinochloa colona*	Poaceae	Jungle rice	Kudiraivalipul
5.	Panicum repens	Poaceae	Torpedo grass	Injipul
	C. Broad Leaved Weeds			
1.	*Acalypha indica*	Euphorbiaceae	Indian acalypha	Kuppai meni
2.	*Ageratum conyzoides*	Compositae	Goat weed	Ageratum
3.	*Amaranthus viridis*	Amaranthaceae	Pig weeds	Kuppai keerai
4.	*Boerhavia diffusa*	Nyctaginaceae	Spreading hug weed	Mukkuratai
5.	*Celosia argentea*	Amaranthaceae	White cock's comb	Pannai keerai
6.	*Cleome viscosa*	Capparidaceae	Wild mustard	Nai kadugu
7.	*Clitoria ternatea*	Leguminosae	Clitoria	Sangu pushpam
8.	*Commelina benghalensis*	Commelinaceae	Day flower	Kannan keerai
9.	*Corchorus olitorius*	Tiliaceae	Jew's mallow	Punnakku poondu
10.	*Corchorus trilocularis*	Tiliaceae	Wild jute	Perumpunnakku poondu
11.	Datura fastuosa/ *D. metel*	Solanaceae	Thorn apple	Umatham
12.	*Digeria arvensis*	Amaranthaceae	Amaranthus	Thoyya keerai
13.	*Digitaria sanguinalis*	Poaceae	Crab grass	Arisipul
14.	*Eclipta alba*	Compositae	False daisy	Karisalanganni
15.	*Gomphrena decumbens*	Amaranthaceae	—	Neer vadamalli
16.	*Gynandropsis pentaphylla*	Capparidaceae	Carvelle seed	Kattukadugu
17.	*Hibiscus micranthus*	Malvaceae	—	Nithya malli
18.	*Hibiscus vitifolicus*	Malvaceae	—	Manja tutter
Sl. No.	**Scientific Name**	**Family**	**Common Name**	**Tamil Name**
19.	*Orabanche cernua*	Oranbanchaceae	Broom rape	Puhaiyilai kalan
20.	*Oxalis corniculata*	Oxalidaceae	The Indian sorrel	Puliyarai keerai
21.	*Portulaca oleracea*	Portulacaceae	Indian purslane	Paruppu keerai
22.	*Portulaca quadrifida*	Portulacaceae	Purslane	Pasalai keerai
23.	*Rhynchosia minima*	Leguminosae	—	Chitt-avarai
24.	*Solanum nigrum*	Solanaceae	Garden night shade	Manathakkali
25.	*Spergula arvensis*	Caryophyllaceae	Corn spurry	Thadi keerai
26.	*Striga lutea*	Scrophularaceae	Witch weed	Pallipoondu
27.	Trianthema decandra	Aizoaceae	Gadabani	Vellai sarani

28.	*Trianthema portulacastrum*	Aizoaceae	Horse purslane	Saranai
29.	*Trianthema triquetia*	Aizoaceae	Horse purslane	Siru saranai
30.	*Vernonia cinerea*	Compositae	Purple fleabane	Mookuthi poondu
31.	Vicoa indica	Compositae	—	Manjamookuthi poondu

4. GENERAL CHARACTERS/ BIOLOGY OF DRYLAND WEEDS

In the drylands of Tamil Nadu, sorghum, groundnut, pulses and cotton are the dominant crops. The detailed survey of the dominant weed flora on farmer's field in dryland areas showed that weeds belonging to 'Poaceae' family are dominant in crops like millets and maize, while broad-leaved weeds are most abundant in pulses and vegetable crops.

Adaptive mechanisms of dryland weeds are:

- Deep root system for effective moisture extraction
- Efficient seed dispersal mechanism
- Long dormancy period
- Modification of above and below ground parts, presence of cuticle, vascular bundles, waxy layers etc.

List of predominant and important weeds of dryland ecosystem

Sl. No.	Scientific Name	Family	Common Name	Tamil Name
1.	*Anagallis arvensis*	Primulaceae	Scarlet pimpernel	—
2.	*Aristolochia bracteata*	Aristolochiaceae	Worm killer	Aduthinna palai
3.	Corchorus trilocularis	Tiliaceae	Wild jute	Perumpunnakku poondu
4.	*Eleusine indica*	Poaceae	Goose grass	Thipparagi
5.	*Eragrostis tenella*	Poaceae	Love grass	—
6.	*Euphorbia hirta*	Euphorbiaceae	Spurge/Garden spurge	Amman pacharisi
7.	Leucas aspera	Labiatae	—	Thumbai
8.	*Leucas utricaefolia*	Labiatae	—	Perum thumbai
9.	*Ocimum canum*	Labiatae	Hoary basil	Naithulasi
10.	*Oldenlandia umbellata*	Rubiaceae	Indian madder	Saya veir
11.	Physalis minima	Solanaceae	Ground cherry/ Sun berry	Thakkali

12.	Sida spinosa	Malvaceae	Prickly sida	Aruvaamookku poondu
13.	*Solanum elaegnifolium*	Solanaceae	White horse nettle	Kattu kandankathiri
14.	*Solanum xanthocarpum*	Solanaceae	Night shade	Kanda kathiri
15.	*Tridax procumbens*	Compositae	Tridax	Vettukaya poondu

5. WEED SURVEY AND WEED VEGETATION ANALYSIS - DENSITY, FREQUENCY, SDR AND IVI

Weed vegetation consists either of the single weed species or more commonly of different weed species that grow in association with crops. In weed control experiments the response of the existing weed vegetation as well as the crop response to various treatments should be evaluated. Knowledge of the various sampling methods employed in weed vegetation analysis will enable the researchers to select the technique that best suits individuals' objectives taking into consideration the resources available.

Need to measure weed vegetation analysis

It is necessary to know the number and nature of weeds competing with the crops. The weed density gives the biological stress that the crops are subjected to. In field study, appropriate weed sampling techniques and weed vegetation analysis are to be employed for arriving useful information on weed vegetation and crop weed competition.

In weed vegetation analysis studies, both destructive and non-destructive methods of sampling are adopted, although they have inherent advantages and disadvantages. However, in destructive method, plants are pulled out of the place or cut at the base and fresh and dry weights are recorded. Dry weight is more reliable estimate than fresh weight since the latter is highly influenced by moisture content of the plants varying across species, growth stages of plants etc. Fresh and dry weight may be more useful to study plant's relative cover since dry weight may correlate with cover. However, a non-destructive methods may be of sampling fosters repeatability of observation and the existing vegetation remains undisturbed and almost unhurt.

It is a common practice to determine the "minimal sample area" of the community, which is defined as the smallest area on which the species composition of the community in question is adequately represented. The smallest area gives an indication of the quadrat size that should be used. The size of the quadrat is decided by "species-area-curve method". Similarly, the number of quadrats required for adequate sampling of all the same of the community is decided.

Weed vegetation analytical parameters are determined by means of three main units *viz.*, area, line and point as employed in quadrat, transect and point methods respectively. A quadrat is frequently used for both destructive and non-destructive sampling and quantitative characters like fresh weight, dry weight, frequency, density, abundance and dominance, cover could be determined. Models of List / Census quadrat to be worked are given below.

The standard weeds vegetation sampling technique

Sl. No.	Sampling technique	Particulars
1.	Quadrat size	0.25 m^2 ($0.5 \times 0.5m$).
2.	Quadrat number	Two quadrat per plot.
3.	Quadrat arrangement	At two locations on opposite sides of the sampling area.
4.	Sampling area	Outside the yield sampling area. Borders of the plot should not be included.
5.	Sampling time	At least at grand growth stages of crop. e.g., maximum tillering and at flowering stages of the rice.

A. Collection of data

Data collected from the quadrat sample should be presented in relation to area. e.g., Density m^{-2}. There are two methods normally adopted for the measurement of weed density.

B. Weed count / Counting method

Total weed count, individual weed species count and group of weeds viz., Grasses, sedges and broad-leaved weeds. e.g., Numbers m^{-2}.

Advantages

Useful method in assessing changes caused by treatments

Disadvantages

Tedious and consumes time and resources, accuracy depends on the size and the number of quadrats

C. Weed weight / Dry matter weight method

Total weed weight, weight of individual weed species and groups of weeds viz., Grasses, sedges and broad-leaved weeds. e.g., kg ha^{-1} or g m^{-2}

Advantages

Precise and more reliable than weed counts and indicates the effect of weed control treatments on the growth of weeds.

Disadvantages

Destruction of plants is required, no reliability, consumes time, resources and manpower.

Precautions

1. The weed count should be taken after the emergence of weeds but at the initial stage of crop growth itself to facilitate the placement of quadrat.
2. Take care to note the name and number of weeds immediately.
3. Select at random, the location (or) site for placing quadrat in different places of the crop field.

Materials Required

- Quadrat of 0.5 m × 0.5m size.
- Observation note book / Pocket note
- Pencil

Procedure

- Select a field with young crop plants and weeds for taking weed count
- Place the quadrat at randomly selected place in the field
- The weeds species are listed and number of individuals of each species is counted
- First pull out all the dicotyledon weeds inside the quadrat taking simultaneous counting
- Record the number of dicotyledon weeds in the observation note book /pocket note
- Then pull out and count all the monocotyledons and record it the observation note book / pocket note
- Multiply the number of monocotyledon weeds by 4 to find out the number of weeds m^2
- In the same way, calculate the number of dicotyledon weeds m^2
- With the help of the above data weed vegetation analysis parameters are calculated as follows

1. Weed Intensity (WI): $\frac{\text{Number of individual species}}{\text{Total number of weed flora}} \times 100$
2. Relative Dry Weight (RDW): $\frac{\text{Dry weight of given weed species}}{\text{Total dry weight of all weed species}}$
3. Absolute Density (D): Total number of individuals of a species in unit area (in m^2)

or

$$\frac{\text{Total number of individuals of a species}}{\text{Total number of quadrats studied}}$$

or

$$\frac{\text{Total number of individuals of a species (in all quadrats)}}{\text{(Total number of quadrats studied)} \times \text{(Area in m}^2\text{of a quadrat)}}$$

4. Relative Density (RD):

$$\frac{\text{Absolute density of the speciesin question}}{\text{Sum/ total of absolute densities of all the species}} \times 100$$

or

$$\frac{\text{Absolute density for a species}}{\text{Total number of plants for all species}} \times 100$$

5. Absolute Frequency (AF):

$$\frac{\text{Number of quadrats in which given species occurred (n)}}{\text{Total number of quadrats studied (N)}} \times 100$$

6. Relative Frequency (RF):

$$\frac{\text{Absolute frequency values of a species}}{\text{Sum / total of absolute frequency values of all species}} \times 100$$

or

$$\frac{\text{Number of occurrence of a species}}{\text{Sum/ total of occurrence of all species}} \times 100$$

7. Dominance (Do):

$$\frac{\text{Absolute density of a species}}{\text{Number of quadrats in which the given species occurred}} \times 100$$

8. Relative Dominance (RDo):

$$\frac{\text{Dominance of a species}}{\text{Sum / total of dominance of all species}} \times 100$$

9. Importance Value Index (IVI): Relative density (RD) + Relative frequency (RE) + Relative dominance (RDo)

10. Summed Dominance Ratio (SDR): IVI / 3 (or) (RD + RE + RDo) / 3

QUESTIONS

I. Fill in the blanks

1. Botanical name of Purple nut sedge ______________
2. Botanical name of Hoorah grass ________________
3. Family of *Cyperus rotundus* is __________________
4. Botanical name of Indian purslane ______________
5. Botanical name of jungle rice __________________

II. Choose the best answers

1. Common name of *Phyllanthus niruri is*
   (a) False daisy (b) Indian acalypha
   (c) Creeping spider ling (d) Gripe weed
2. Which weed is comes under grasses
   (a) *Dactyloctenium aegyptium* (b) *Panicum repens*
   (c) Both A & b (d) *Commelina benghalensis*
3. Scientific name of Thorn apple
   (a) *Datura fastuosa* (b) *Rhynchosia minima*
   (c) *Digitaria sanguinalis* (d) None of the above
4. Common name of *Cynodon dactylon*
   (a) Finger grass (b) Bermuda grass
   (c) Torpedo grass (d) Crow foot grass

III. Answer the following questions

1. Write biology of dryland and wet land weeds with example
2. Write character of garden land weeds with example
3. List out four wet land, gardend and dryland land weeds with botanical name

15

CLASSIFICATION AND CHARACTERISTICS OF WEEDS - AQUATIC, PARASITIC AND OBNOXIOUS WEEDS

Aquatic weeds are unwanted plants, growing in or near aquatic bodies such as canals, ponds, lakes, rivers, reservoirs etc.

1. IMPORTANT AQUATIC WEEDS OF INDIA

Over 140 species of aquatic weeds have been reported in India, but the most important and problematic ones are *Eichhornea crassipes* (water hyacinth) *Salvinia molesta* (water fern) *Salvinia natans* (Water fern) *Nymphaea odorata* (water lily), *Nelumbo lutea* (lotus) *Hydrilla verticillata* (hydrilla), V*alisneria spiralis* (tape grass or eel weed) *Typha angustata* (cat tail) *Nympholdes, Potamogeton spp*. (pond weed) *pistia stratiotes* (water lettuce) and algal weeds like Chara sp. *Nitella sp*. and bloom-farming blue-green algae

2. KINDS OF PROBLEMS CAUSED BY AQUATIC WEEDS

Increase the loss of water by evapo-transpiration the floating weeds such as *Eichhornia crassipes* and *Salvinia* are said to increase loss of water 2-4 times more than the exposed water surface. Similarly the emerged weeds, such as *Typha spp* and rushes increase water losses by 150-200 percent.

Reduce the flow-rate of water in canals, channels drains, rivers etc., and thus increase loss of water by seepage and evaporation and cause more silting in canal beds. In canals severely infested with aquatic weeds, the flow-rate can reduce to 80-90 percent.

Hinder pisce culture Aquatic weeds obstruct free-movement of fish, and adversely affect the growth of planktons, the primary food of fish.

Interfere with hydro-electric projects and irrigation schemes The aquatic weeds choke the turbines and gates of water supply, small aquatic weeds such as Pistia, and Lemna choke the siphons, sprinklers and intake of water pumps used for irrigation.

Interfere with movement of boats, for transport, recreation and fishing.

Create problem for lowland paddy cultivation floating aquatic weeds like, hyacinth S*alvinia and pistia* etc. enter the paddy fields and suppress and compete with crop plants.

Provide habitat for the development of disease-causing insects, such as mosquitoes which spread diseases like malaria, dengue fever, filariasis and encephalitis.

Cause silting of reservoirs lakes etc., because vigorously grown floating and submerged weeds die decompose and settle at the bottom as sediments.

Aquatic weeds reducing the value of lakes, tanks, reservoirs streams etc.

3. CLASSIFICATION OF AQUATIC WEEDS

The aquatic weeds are generally classified in the following manner

3.1 Algae

(a) Planktonic or Phytoplanktonic

(b) Filamentous algae

3.1.1 Algae are simple photo-synthetic plants

Having no vascular tissues nor they differentiated into stem, roots or leaves. All are aquatic or semi-aquatic plants.

3.1.2 Phytoplankton algae

Autotrophic, prepare their own food, compete with fish for O2, and release CO2, These are single celled, filamentous or colonial forms. Heavy growth of algae may give the colour to the water body green, yellow, red or black. They form water blooms or scum. They are source of food for many aquatic fish. Excessive phytoplantonic blooms often results in zoo planktonic development that may deplete the oxygen, eutrophication and destruction of fish and other aquatic wild life. Important Genera are: BGA.

3.1.3 Filamentous algae or non planktonic

These are single celled, joined end to end, which may form as a single thread, branched filaments, nets or forked leaf like forms. These are capable of changing the odour and quality of water, they not only infest domestic water but also industrial water. *Chara* and *Nitella*, *Spirogyra and Cladophora.*

3.2 Vascular hydrophtes

Over 100 families of vascular hydrophytes are known. These are the plants which do not have strong stem and conducting tissue, not well developed as compared to mesophyts.

They are classified as under

3.2.1. Emerged weeds (Surface weeds)

Emerged weeds, which have roots in the bottom soil but the leaves and flowers emerge to water surface

Examples: *Nelumbo lutea* (lotus),*Nymphaeo odorata* (water lily),*Trapa natans* (water chestnut).and Nymphoides (Floating heart.)

3.2.2 Submersed weeds

Submerged weeds are those that remain below water surface.

Most of the vegetative growth of these weeds takes place beneath the water surface. They are further differentiated into:

3.2.3 Rooted submersed

These weeds have roots, Examples: *Hydrilla verticillata* (hydrilla),*Valisneria spiralis* (tape grass or eel weed),*Potamogeton spp*. (pond weed).

3.2.4 Suspended submersed

These weeds have no roots. *Ceratophyllum demersum* (coon tail) and *Utricularia* bladderwort are examples:

3.3 Floating Weeds

Which remain free floating with their leaves above surface and roots within water. Their leaves floating on the water surface either singly or in rosettes. They have true roots and leaves.

3.3.1 Free floating

Eichornia, Pistia, Salvinia

3.3.2 Rooted floating

Nymphaea, Pistia Salvinia

3.4 Marginal weeds

Weeds can grow on saturated soil above the water surface. They can grow from moist shore line area into water upto 60-90cm depth. Eg. *Typha, Polygonum, Alternanthera, Ipomoea etc.*

4. MANAGEMENT OF AQUATIC WEEDS

Many aquatic weeds are desirable for fish and other aquatic fauna, but their rapid and excessive growth poses various problems. Eradication of aquatic weeds is therefore desirable only in certain situations. Generally only their excessive growth is to be contained.

4.1 Preventive measures

These measures involve,

- Prohibiting their introduction from one area to another within the country by legislative measures.
- Monitoring major water bodies once cleared from infestation to prevent re-infestation.

- Installation of sieves at the entry points of water from canals into paddy fields, to prevent infestation of free-floating weeds like water-hyacinth.
- Desiccation of ponds and tanks heavily infested with aquatic weeds by draining out water during summer.

4.2 Mechanical control measures

Manual removal, cutting, chaining, dredging, netting, picking by machine, mowing burning and some of the mechanical methods to deal with aquatic weeds.

4.2 Manual removal

When infestation is limited manual removal using rakes and pitch-forks for floating weeds and sickle and scythes for ditchbank weeds and emerged static weeds is feasible.

Cutting: For submerged and emerged static weeds under-water weed cutters mounted on motor boats are used to cut and remove the weeds.

Chaining: Dragging of heavy chain in canals, rivulets or lakes by means of a pair of tractors or boats (as the situation may be) to break submersed weeds into fragments.

Dredging: Dredging is the total removal of weeds from bottom, along with mud, which is a very effective means to control tuberous rhizomatous weeds.

Netting: Small floating weeds like duck-weed (Lemma) Pistia, Azolla can be removed from fishponds by using dragnets.

Weed harvesters: Now-a-days mechanical weed harvesters launched on slow-moving boats are available. These machines can pick-up floating weeds and throw them on to the shore.

4.3. Chemical control

Control of aquatic weeds with herbicides is efficient, easier, faster and even less expensive. However in a country like India, where water resources are used for various purposes, herbicides application could prove hazardous to human beings, cattle, fish and other useful aquatic fauna and non-target plants.

After chemical application usually there should be minimum 7 days waiting period for using water for various purposes.Under Indian situations the following herbicides have been recommended to control various types of aquatic weeds.

Chemicals

- Acrolin: Effective against submerged weeds. It is applied in a concentration ranging from 4 ppm. Eg. *Potamogeton*
- Dalapon: Effective against grasses and cat tail @15-20kg/ha
- Dalapon 5-10kg/ha + amitrole T @2kg/ha are effective against cat tails.
- Amitrole is suitable for cat tails.
- Amitrole T suitable for water hyacinth, alligator weed and other emerged weeds @ 5-10kg/ha.

- 2,4-D @ 4-8kg/ha effective against floating and emerged weeds.

  Note: Fish from water treated with 2,4-D and 2,4,5-T and silvex should not be used for consumption before the 30 days of the treatment.
- Diquat and Paraquat @ 0.5-1.0 ppm control submerged and algae weeds. They should be applied before they reach the surface of the water.
- Endothal @ 1-5ppm is very useful for controlling many submerged weeds.
- Fenac @ 15-20kg/ha is effective in preventing regeneration of roots and rhizomes of aquatic weeds (emerged aquatic weeds).
- Simazine, diuron and monuron are used as algicide @1-2ppm.
- Copper sulfate is effective against all kinds of algae including chara.

QUESTIONS

I. Fill in the blanks

1. Scientific name of water hyacinth is ________
2. Aquatic weeds reduce the canal flow-rate of up to ________percent.
3. *Nelumbo lutea is* ________ aquatic weeds
4. Vegetative growth of weeds beneath the water surface is called ________
5. ________ is rooted submersed weed

II. Choose the best answers

1. Significant examples for emerged aquatic weeds
   (a) *Lotus* (b) *Water lilly* (c) *Hydrilla* (d) Both A & B
2. Mechanical control measures of aquatic weeds
   (a) Cutting (b) Chaining (c) Netting (d) All the above
3. Chemical recommend for control floating and emerged weeds
   (a) Dalapon (b) Amitrole
   (c) Diquat and Paraquat (d) 2,4-D
4. Chemical used for *Algicide*
   (a) Copper sulfate (b) diuron (c) Simazine (d) *All the above*

III. Answer the following questions

1. Write the classification of aquatic weeds and their characteristics
2. Write the problems caused by aquatic weeds
3. Different method of control measures of aquatic weeds

16

CROP WEED INTERACTIONS - CRITICAL CROP WEED COMPETITION, COMPETITIVE AND ALLELOPATHIC EFFECTS OF WEEDS AND CROPS

1. CROP WEED INTERACTIONS

Competition and allelopathy are the main interactions, which are of importance between crop and weed. Allelopathy is distinguished from competition because it depends on a chemical compound being added to the environment while competition involves removal or reduction of an essential factor or factors from the environment, which would have been otherwise utilized.

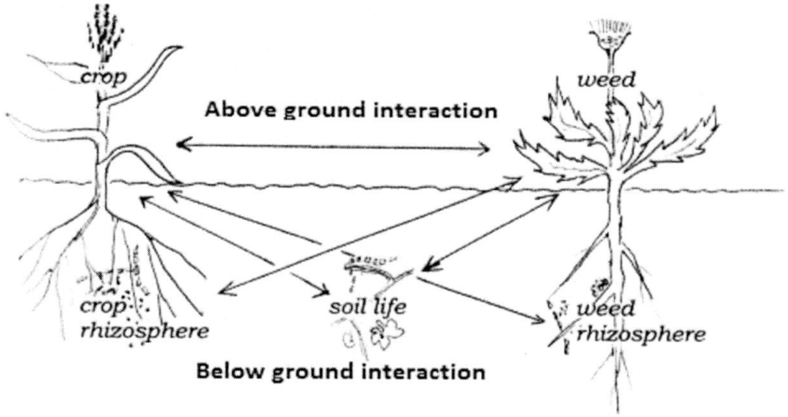

2. CRITICAL PERIOD OF WEED COMPETITION

Critical period of weed competition is defined as the shortest time span during the crop growth when weeding results in highest economic returns.

The critical Period of crop-weed competition is the period from the time of sowing up to, which the crop is to be maintained in a weed free environment to get the highest economical yield. The weed competition in crop field is invariably severe in early stages of crop than at later stages. Generally in a crop of 100 days duration, the first 35 days after sowing should be maintained in a weed free condition. There is no need to attempt for a weed free condition throughout the life period of the crop, as it will entail unnecessary additional expenditure without proportionate increase in yield.

Critical crop weed competition period for important field crops

S. No.	Crop		Critical period of weed competition
1.	Rice Transplanted Direct sowing	 : :	 Up to 6 WAT Up to 6 WAS
2.	Wheat and Ragi	:	Up to 5 WAS
3.	Maize	:	2 to 6 WAS
4.	Bajra	:	Up to 4 WAS
5.	Jowar and Sesame	:	2-5 WAS
6.	Sugarcane	:	12 to 14 WAS
7.	Sunflower	:	2 to 4 WAS
8.	Groundnut	:	4 to 7 WAS
9.	Green gram and Black gram	:	4 to 6 WAS
10.	Red gram	:	Up to 7 WAS
11.	Horse gram	:	3 to 6 WAS
12.	Cotton	:	6 to 9 WAS
13.	Castor	:	3 to 9 WAS
14.	Tobacco	:	Up to 9 WAT

3. FACTORS AFFECTING CROP WEED COMPETITION

3.1 Period of weed growth

It is important to note that when the crop-weed competition is critical; when the weed must be removed; when they may be allowed to grow without fear of any appreciable crop damage. The weeds that germinate and grow at the same time as the crop seeds are most successful competitors. If weeds are controlled at early period of crop growth, the optimum crop yield could be harvested which may be quite close to the maximum

3.2 Intensity of weeds

In general increasing the weed intensity results in reducing the growth and yield of crops.

3.3 Weed species

Different weeds differ in their ability to compete with crop at similar density levels. This is primarily because of difference in their growth habits as well as due to the allelopathic affects which they may exert on the germination and growth for the crop

3.4 Soil fertility

Since weeds compete with corps for mineral nutrients, particularly nitrogen. Addition of fertilizers, manures and other amendments will lessen weed-stress on crop. Methods and time of application of fertilizer to the crop fields are important in determining whether the added fertilizers will suppress or increase weed growth in fields.

3.5 Soil moisture status

Higher water use efficiency of the crops has been noted in weed free plots than in the weedy plots because of higher level of crop yield in the weed free plots. Time of rain or first irrigation after planting of the crops is an important factor determining the intensity of the crop weed competition

3.6 Soil reaction

Some of the weed sp. tolerates abnormal soil reactions much better than others. Some weeds grow vigorously on abnormal soil pH, where majority of our crop plant will be stunted in growth. Weed crop competition is more intense on high or low pH soils than on normal pH soils.

3.7 Cropping practices

- Date and method of planting of crop
- Crop density and rectangularity
- Crop spp. and varieties.

4. COMPETITIVE EFFECTS OF WEEDS ON CROPS

4.1 Competition for Nutrients

Weeds usually absorb mineral nutrients faster than many crop plants and accumulate them in their tissues in relatively larger amounts.

- The associated weed is responsive to nitrogen and it utilizes more of the applied 'N' than the crop. Eg. The 'N' uptake by *Echinochloa crusgalli* is more than rice.
- Nutrient removal by weeds leads to huge loss of nutrients in each crop season, which is often twice that of crop plants. For instance at early stages of maize cultivation, the weeds found to remove 9 times more of N, 10 times more of P and 7 times more of K. The NPK uptake of major weeds and crops are given below

S. No.	Species	N	P	K
Weeds				
1	Achyranthus aspera	2.21	1.63	1.32
2	Amaranthus viridis	3.16	0.06	4.51
3	Chenapodium album	2.59	0.37	4.34

4	Cynodon dactylan	1.72	0.25	1.75
5	Cyperus rotundus	2.17	0.26	2.73
Crops				
1	Rice	1.13	0.34	1.10
2	Sugarcane	0.33	0.19	0.67
3	Wheat	1.33	0.59	1.44

4.2 Competition for moisture

- In general, for producing equal amounts of dry matter, weeds transpire more water than do most of our crop plants. It becomes increasingly critical with increasing soil moisture stress, as found in arid and semi-arid areas.
- As a rule, C_4 plants utilize water more efficiently resulting in more biomass per unit of water. *Cynodon dactylon* had almost twice as high transpiration rate as pearl millet.
- In weedy fields soil moisture may be exhausted by the time the crop reaches the fruiting stage, i.e. the peak consumptive use period of the crop, causing significant loss in crop yields.

4.3 Competition for light

- It may commence very early in the cop season if a dense weed growth smothers the crop seedlings.
- It becomes important element of crop-weed competition when moisture and nutrients are plentiful.
- In dry land agriculture in years of normal rainfall the crop-weed competition is limited to nitrogen and light.
- Unlike competition for nutrients and moisture once weeds shade a crop plant, increased light intensity cannot benefit it.

4.4 Competition for space (CO_2)

Crop-weed competition for space is the requirement for CO_2 and the competition may occur under extremely crowded plant community condition. A more efficient utilization of CO_2 by C_4 type weeds may contribute to their rapid growth over C_3 type of crops.

5. EFFECT OF WEED COMPETITION ON CROP GROWTH AND YIELD

1. Crop growth and yield is affected
2. Crop suffers from nutritional deficiency
3. Leaf area development is reduced
4. Yield attributes will be lowered
5. Reduce the water use by the crop
6. Affect the dry matter production

7. Lowers the input response
8. Causes yield reduction
9. Pest and disease incidence will be more

6. LOSSES CAUSED BY WEEDS

6.1 Reduction in crop yield

Weeds compete with crop plants for nutrients, soil moisture, space and sunlight. In general an increase in one kilogram weed growth corresponds to reduction in one kilogram of crop growth. Among the pests weeds account for 45 % reduction in yield while the insects 30%, diseases 20% and other pests 5%. The table below depicts the percentage range of yield loss due to weeds in some important field crops.

S. No.	Crop	Yield losses (%)
1.	Rice	9.1 – 51.4
2.	Wheat	6.3 – 34.8
3.	Maize	29.5 – 74.0
4.	Millets	6.2 – 81.9
5.	Groundnut	29.7 – 32.9
6.	Sugarcane	14.1 – 71.7
7.	Linseed	30.9 – 39.1
8.	Cotton	20.7 – 61.0
9.	Carrot	70.2 – 78.0
10.	Peas	25.3 – 35.5

6.2 Loss in crop quality

If a crop contains weed seeds it is to be rejected, especially when the crop is grown for seed. For example, the wild oat weed seeds are similar in size and shape of the crops like barley, wheat, and its admixture may lead to rejection for seed purpose.

6.3 Weeds as reservoirs of pests and diseases

Weeds form a part of community of organisms in a given area. Consequently, they are food sources for some animals, and are themselves susceptible to many pests and diseases.

6.4 Interference in crop handling

Some weeds can make the operation of agricultural machinery more difficult, more costly and even impossible. Heavy infestation of *Cynadon dactylon* causes poor ploughing performance.

6.5 Reduction in land value

Heavy infestation by perennial weeds could make the land unsuitable are less suitable for cultivation resulting in loss in its monetary value.

6.6 Limitation of crop choice

When certain weeds are heavily infested, it will limit the growth of a particular crop. The high infestation of parasitic weeds such as *Striga lutea* may limit the growing of sorghum or sugarcane.

6.7 Loss of human efficiency

Weeds reduce human efficiency through physical discomfort caused by allergies and poisoning. Weeds such as congress weed (*Parthenium hysterophorus*) causes itching. Thorny weeds like *Solanum* spp. restrict moment of farm workers in carrying out farm practices such as fertilizer application, insect and disease control measures, irrigation, harvesting etc.

6.8 Problems due to aquatic weeds

The aquatic weeds that grow along the irrigation canals, channels and streams restricts the flow of water.

7. FACTORS AFFECTING THE COMPETITIVE ABILITY OF CROPS AGAINST WEEDS

(a) Density of weeds
(b) Crop density
(c) Type of weeds species
(d) Type of crop species and their varieties
(e) Soil factors
(f) Climate
(g) Time of germination
(h) Cropping practices
(i) Crop maturity

8. ALLELOPATHY

Allelopathy is the detrimental effects of chemicals or exudates produced by one (living) plant species on the germination, growth or development of another plant species (or even microorganisms) sharing the same habitat.

Allelopathy does not form any aspect of crop-weed competition, rather, it causes Crop-Weed interference, it includes competition as well as possible allelopathy.

Allelo chemicals are produced by plants as end products, by-products and metabolites liberalised from the plants; they belong to phenolic acids, flavanoides, and other aromatic compounds viz., terpenoids, steroids, alkaloids and organic cyanides.

8.1 Allelopathic effect of weeds on crops

Maize

- Leaves & inflorescence of *Parthenium* sp. affect the germination and seedling growth
- Tubers of *Cyperus esculentus* affect the dry matter production

Sorghum

- Stem of *Solanum* affects germination and seedling growth
- Leaves and inflorescence of *Parthenium* affect germination and seedling growth

Wheat

- Seeds of wild oat affect germination and early seedling growth
- Leaves of *Parthenium* affects general growth
- Tubers of *C. rotundus* affect dry matter production
- Green and dried leaves of *Argemone mexicana* affect germination & seedling growth

Sunflower

- Seeds of *Datura* affect germination & growth

8.2 Allelopathic effect of crop plants on weeds

(i) Root exudation of maize inhibits the growth of *Chenopodium album*

(ii) The cold water extracts of wheat straw when applied to weeds reduce germination and growth of *Abutilon* sp.

8.3 Allelopathic effect of weeds on weeds

- Extract of leaf leachate of decaying leaves of *Polygonum* contains flavonoides which are toxic to germination, root and hypocotyls growth of weeds like *Amaranthus spinosus*
- Inhibitor secreted by decaying rhizomes of *Sorghum halepense* affect the growth of *Digitaria sanguinalis* and *Amaranthus* sp.

9. FACTORS INFLUENCING ALLELOPATHY

9.1 Plant factors

(i) Plant density: Higher the crop density the lesser will be the allelo chemicals it encounters

(ii) Life cycle: If weed emerges later there will be less problem of allelochemicals

(iii) **Plant age:** The release of allelochemicals occurs only at critical stage. For eg. in case of *Parthenium*, allelopathy occurs during its rosette & flowering stage.

(iv) **Plant habit:** The allelopathic interference is higher in perennial weeds.

(v) **Plant habitat:** Cultivated soil has higher values of allelopathy than uncultivated soil.

9.2 Climatic factors

The soil & air temperature as well as soil moisture influence the allelo chemicals potential

9.3 Soil factors

Physico-chemical and biological properties influence the presence of allelochemicals.

9.4 Stress factors

Abiotic and Biotic stresses may also influence the activity of allelochemcals

10. MECHANISM OF ACTION OF ALLELOCHEMICALS

- Interfere with cell elongation
- Interfere with photosynthesis
- Interfere with respiration
- Interfere with mineral ion uptake
- Interfere with protein and nucleic acid metabolism

Use of allelopathy in biological control of weeds

1. Use of cover crop for biological control
2. Use of alleopathic chemicals as bio-herbicides

QUESTIONS

I. Fill in the blanks

1. Generally in a crop of 100 days duration, the first _____ days after sowing should be maintained in a weed free condition.
2. Critical crop weed competition period for maize is ___ to ____ WAS
3. At early stages of maize cultivation, the weeds found to remove 10 times more of ______.
4. CO_2 utilization efficiency is higher in weeds due to _________ nature

II. Choose the best answers

1. Critical crop weed competition period for rice crop is ___________
   (a) Upto 4 WAT (b) Upto 6 WAT
   (c) Entire crop period (d) Upto 7 WAT

2. Critical crop weed competition period for sugarcane is ___________
   (a) 7 WAS (b) 12 to 15 WAS
   (c) 20 WAS (d) None of these
3. *Echinochloa crusgalli* utilize more _______ than rice crop
   (a) Nitrogen (b) Phosphorus (c) Zinc (d) Megnesium
4. __________had almost twice as high transpiration rate as pearl millet.
   (a) *Cynodon dactylon* (b) *Amaranthus*
   (c) *Cyperus* (d) All the above
5. Yield losses in millet due to weed infestation is _________
   (a) 6 to 8 % (b) 15 (c) 45 % (d) 12 %
6. Leaves and inflorescence of ___________ affect the germination and seedling growth
   (c) *Parthenium* sp. (b) *Cyperus sp.*
   (c) *Solanum sp.* (d) *Phyllanthus sp.*

III. Answer the following questions

1. Crop weed interactions
2. Critical period of crop weed competition
3. Why the weeds are being good competitor for crops?
4. Allelopathy and types
5. Effect of weed competition on crop growth and yield
6. Yield losses of major field crops by weeds

17

PRINCIPLES AND METHODS OF WEED MANAGEMENT: PREVENTIVE, CULTURAL, MECHANICAL, CHEMICAL, BIOLOGICAL AND ALTERNATE METHODS

Principles of weed management are,

(a) Prevention
(b) Eradication
(c) Weed control
(d) Weed management

1. PREVENTIVE

(i) Sowing of weed free clean seed
(ii) Use of clean implements
(iii) Removal of weed along canal and irrigation channel
(iv) Care in transplanting of seedlings/plantlets
(v) Use of well rotten manure
(vi) Avoiding passing of cattle through weed infested area
(vii) Crop management practices
(viii) Enforcement of weed Laws
(ix) Quarantine Laws
(x) Use of pre-emergence herbicides

2. ERADICATION: (IDEAL WEED CONTROL RARELY ACHIEVED)

It infers that a given weed species, its seed & vegetative part has been killed or completely removed from a given area & that weed will not reappear unless reintroduced to the area. Because of its difficulty & high cost, eradication is usually attempted only in smaller areas such as few ha., a few thousand m^2 or less. Eradication is often used in high value areas such as green houses, ornamental plant beds & containers. This may be desirable and economical when the weed species is extremely noxious and persistent as to make cropping difficult and economical.

3. WEED CONTROL

It encompasses those processes where by weed infestations are reduced but not necessarily eliminated. It is a matter of degree ranging from poor to excellent. In control methods, the weeds are seldom killed but their growth is severely restricted, the crop makes a normal yield. In general, the degree of weed control obtained is dependent on the characters of weeds involved and the effectiveness of the control method used.

4. WEED MANAGEMENT

Weed control aims at only putting down the weeds present by some kind of physical or chemical means while weed management is a system approach whereby whole land use planning is done in advance to minimize the very invsasion of weeds in aggressive forms and give crop plants a strongly competitive advantage over the weeds.

5. WEED CONTROL

Weed control methods are grouped into cultural, physical/mechanical, chemical and biological. Every method of weed control has its own advantages and disadvantages. No single method is successful under all weed situations. Many a time, a combination of these methods gives effective and economic control than a single method.

5.1 Cultural weed control

Several cultural practices like tillage, planting, fertiliser application, irrigation etc., are employed for creating favourable condition for the crop. These practices if used properly, help in controlling weeds. Cultural methods, alone cannot control weeds, but help in reducing weed population. They should, therefore, be used in combination with other methods. In cultural methods, tillage, fertiliser application. and irrigation are important. In addition, aspects like selection of variety, time of sowing, cropping system, cleanliness of the farm etc., are also useful in controlling weeds.

5.1.1 Field preparation

The field has to be kept weed free. Flowering of weeds should not be allowed. This helps in prevention of build up of weed seed population.

5.1.2 Summer tillage

The practice of summer tillage or off-season tillage is one of the effective cultural methods to check the growth of perennial weed population in crop cultivation. Initial tillage before cropping should encourage clod formation. These clods, which have the weed propagules, upon drying desiccate the same. Subsequent tillage operations should break the clods into small units to further expose the shriveled weeds to the hot sun.

5.1.3 Maintenance of optimum plant population

Lack of adequate plant population is prone to heavy weed infestation, which becomes, difficult to control later. Therefore practices like selection of proper seed, right method of sowing, adequate seed rate protection of seed from soil borne pests and diseases etc. are very important to obtain proper and uniform crop stand capable of offering competition to the weeds.

5.1.4 Crop rotation

The possibilities of a certain weed species or group of species occurring is greater if the same crop is grown year after year. In many instances, crop rotation can eliminate at least reduce difficult weed problems. The obnoxious weeds like *Cyperus rotundus* can be controlled effectively by including low land rice in crop rotation.

Crop rotation is a planned sequence of crops growing in the same field year after year. Rotating crops adds diversity to the cropping system, increasing the sustainability of the system. Although crop rotations are sometimes planned years in advance, they are also dynamic; farmers adapt their crop rotations in response to environmental, pest and market forces.

Crop rotation provides the foundation for long-term weed management. Planting a wide variety of crops with varied characteristics reduces the likelihood that specific weed species will become adapted to the system and become problematic. The success of rotation systems for weed suppression appears to be based on the use of crop sequences that employ varying patterns of resource competition, allelopathic interference, soil disturbance, and mechanical damage to provide an unstable and frequently inhospitable environment that prevents the proliferation of a particular weed species.

Rotating crops allows farmers to rotate other aspects of the crop management system including timing and type of tillage operation, seeding date, timing and type of herbicide application, and type and amount of fertilizer application. Some of the crop characteristics to consider when planning a diverse rotation include:

- grassy vs. broadleaved
- annual, biennial, or perennial
- cool season vs. warm season
- seeding date
- growth habit
- competitive ability
- fertility requirements

Crop rotation can be used to set the stage for eliminating herbicides from one growing season. Farmers participating in an on-farm research project on Pesticide Free Production reported crop rotation as one of the most important tools for preparing for a pesticide-free crop (Nazarko et al., 2003).

5.1.5 Growing of intercrops

Inter cropping suppresses weeds better than sole cropping and thus provides an opportunity to utilize crops themselves as tools of weed management. Many short duration pulses viz., green gram and soybean effectively smother weeds without causing reduction in the yield of main crop.

5.1.6 Mulching

Mulch is a protective covering of material maintained on soil surface. Mulching has smothering effect on weed control by excluding light from the photosynthetic portions of a plant and thus inhibiting the top growth. It is very effective against annual weeds and some perennial weeds like *Cynodon dactylon.* Mulching is done by dry or green crop residues, plastic sheet or polythene film. To be effective the mulch should be thick enough to prevent light transmission and eliminate photosynthesis.

5.1.7 Solarisation

This is another method of utilisation of solar energy for the desiccation of weeds. In this method, the soil temperature is further raised by 5 – 10 ºC by covering a pre-soaked fallow field with thin transparent plastic sheet. The plastic sheet checks the long wave back radiation from the soil and prevents loss of energy by hindering moisture evaporation.

5.1.8 Stale seedbed

A stale seedbed is one where initial one or two flushes of weeds are destroyed before planting of a crop. This is achieved by soaking a well prepared field with either irrigation or rain and allowing the weeds to germinate. At this stage a shallow tillage or non- residual herbicide like paraquat may be used to destroy the dense flush of young weed seedlings. This may be followed immediately by sowing. This technique allows the crop to germinate in almost weed-free environment.

5.1.9. Blind tillage: The tillage of the soil after sowing a crop before the crop plantsemerge is known as blind tillage. It is extensively employed to minimise weed intensity in drill sowing crops where emergence of crop seedling is hindered by soil crust formed on receipt of rain or irrigation immediately after sowing.

5.1.10 Crop management practices

Good crop management practices that play an important role in weed control are

(i) Vigorous and fast growing crop varieties are better competitors with weeds.
(ii) Proper placement of fertilizers ensures greater availability of nutrients to crop plants, thus keeping the weeds at a disadvantage.
(iii) Better irrigation practices to have a good head start over the weeds
(iv) Proper crop rotation programme
(v) Higher plant population per unit area results in smothering effect on weed growth

Merits of cultural method

1. Low cost for weed control
2. Easy to adopt

3. No residual Problem
4. Technical skill is not involved
5. No damage to crops
6. Effective weed control
7. Crop-weed ecosystem is maintained

Demerits of cultural method

1. Immediate and quick weed control is not possible
2. Weeds are kept under suppressed condition
3. Perennial and problematic weeds can not be controlled
4. Practical difficulty in adoption

5.2. Mechanical / physical weed control

Mechanical or physical methods of weed control are being employed ever since man began to grow crops. The mechanical methods include tillage, hoeing, hand weeding, digging cheeling, sickling, mowing, burning, flooding, mulching etc.

5.2.1 Tillage

Tillage removes weeds from the soil resulting in their death. It may weaken plants through injury of root and stem pruning, reducing their competitiveness or regenerative capacity. Tillage also buries weeds. Tillage operation includes ploughing, discing, harrowing and leveling which is used to promote the germination of weeds through soil turnover and exposure of seeds to sunlight, which can be destroyed effectively later. In case of perennials, both top and underground growth is injured and destroyed by tillage.

5.2.2 Hoeing

Hoe has been the most appropriate and widely used weeding tool for centuries. It is however, still a very useful implement to obtain results effectively and cheaply. It supplements the cultivator in row crops. Hoeing is particularly more effective on annuals and biennials as weed growth can be completely destroyed. In case of perennials, it destroyed the top growth with little effect on underground plant parts resulting in re-growth.

5.2.3 Hand weeding

It is done by physical removal or pulling out of weeds by hand or removal by implements called khurpi, which resembles sickle. It is probably the oldest method of controlling weeds and it is still a practical and efficient method of eliminating weeds in cropped and non-cropped lands. It is very effective against annuals, biennials and controls only upper portions of perennials.

5.2.4 Digging

Digging is very useful in the case of perennial weeds to remove the underground propagating parts of weeds from the deeper layer of the soil.

5.2.5 Cheeling

It is done by hand using a cheel hoe, similar to a spade with a long handle. It cuts and shapes the above ground weed growth.

5.2.6 Sickling and mowing

Sickling is also done by hand with the help of sickle to remove the top growth of weeds to prevent seed production and to starve the underground parts. It is popular in sloppy areas where only the tall weed growth is sickled leaving the root system to hold the soil in place to prevent soil erosion. **Mowing** is a machine-operated practice mostly done on roadsides and in lawns.

5.2.7 Burning

Burning or fire is often an economical and practical means of controlling weeds. It is used to (a) dispose of vegetation (b) destroy dry tops of weeds that have matured (c) kill green weed growth in situations where cultivations and other common methods are impracticable.

5.2.8 Flooding

Flooding is successful against weed species sensitive to longer periods of submergence in water. Flooding kills plants by reducing oxygen availability for plant growth. The success of flooding depends upon complete submergence of weeds for longer periods.

Merits of mechanical method

1. Oldest, effective and economical method
2. Large area can be covered in shorter time
3. Safe method for environment
4. Does not involve any skill
5. Weeding is possible in between plants
6. Deep rooted weeds can be controlled effectively

Demerits of mechanical method

1. Labour consuming
2. Possibility of damaging crop
3. Requires ideal and optimum specific condition

5.3. Chemical weed control

Using chemicals, generally referred as herbicides, for the control of weeds is called chemical weed control. In 1944 - discovery of 2,4-D Na salt as a land mark in herbicide usage.

5.3.1. Principles of chemical weed control

The selectivity exhibited by certain chemicals to cultivated crops in controlling its associated weeds without affecting the crops forms basis for the chemical weed control. Such selectivity may be due to differences in the morphology, differential absorption, differential translocation, differential deactivation etc.

Merits

1. Herbicide can be recommended for adverse soil and climatic conditions, as manual weeding is highly impossible during monsoon season.
2. Herbicide can control weeds even before they emerge from the soil so that crops can germinate and grow in completely weed-free environment at early stages. It is usually not possible with physical weed comfort.
3. Weeds, which resemble like crop in vegetative phase may escape in manual weeding. However, these weeds are controlled by herbicides.
4. Herbicide is highly suitable for broadcasted and closely spaced crops.
5. Controls the weeds without any injury to the root system of the associated standing crop especially in plantationcrops like Tea and Coffee.
6. Controls many perennial weed species
7. Herbicides control the weed in the field itself or *insitu* controlling where as mechanical method may lead to dispersal of weed species through seed
8. It is profitable where labour is scarce and expensive
9. Highly economical

Demerits

1. Pollutes the environment
2. Affects the soil microbes if the dose exceeds
3. Herbicide causes drift effect to the adjoining field
4. It requires certain amount of minimum technical knowledge for calibration
5. Leaves residual effects
6. Some herbicide is highly costlier
7. Suitable herbicides are not available for mixed and inter-cropping system.

5.4 Biological and alternate methods of weed control

Use of living organisms viz., insects, disease organisms, herbivorous fish, snails or even competitive plants for the control of weeds is called biological control. In biological control method, it is not possible to eradicate weeds but weed population can be reduced. This method is not useful to control all types of weeds. Introduced weeds are best targets for biological control.

5.4.1. Outstanding and feasible examples of biological weed control

(a) Larvae of *Coctoblastis cactorum*, a moth borer, control prickly pear *Opuntia* sp. The larvae tunnel through the plants and destroy it. In India it is controlled by cochinial insects *Dactylopius indicus* and *D. tomentosus*

(b) *Lantana camara* is controlled by larvae of C*rocidosema lantana*, a moth bores into the flower, stems, eat flowers and fruits.

(c) *Cuscuta* spp. is controlled by M*elanagromyza cuscutae*

(d) *Cyperus rotundus* - *Bactra verutana* a moth borer

(e) *Ludiwigia parviflora* is completely denuded by *Altica cynanea* (steel blue beetle)

(f) Herbivorous fish Tilapia controls algae. Common carp, a non-herbivorous fish controls sub-mersed aquatic weeds. It is apparently due to uprooting of plants while in search of food. Snails prefer submersed weeds.

5.4.2 Bio-Herbicides / Mycoherbicides

The use of plant pathogen which are expected to kill the targeted weeds. A bioherbicide based on a fungus is called a mycoherbicide. A bioherbicide is a preparation of living inoculum of a plant pathogen, formulated, and applied in a manner analogous to that of a chemical herbicide in an effort to control or suppress the growth of weed species.

No	Product	Content	Target weed
1	Devine	A liquid suspension of fungal spores of *Phytophthora palmivora* causes root rot.	Strangle vine (*Morrenia odorata*) in citrus
2	Collego	Wettable powder containing fungal spores of *Colletotrichum gloeosporoides* causes stem and leaf blight	Joint vetch (*Aeschyomone virginica*) in rice, soybean
3	Bipolaris	A suspension of fungal spores of *Bipolaris sorghicola*	Jhonson grass (*Sorghum halepense*)
4	Biolophos	A microbial toxin produced as fermentation product of *Steptomyces hygroscopicus*	Non-specific, general vegetation

QUESTIONS

I. Fill in the blanks

1. High soil temperature can suppress weed seed germination and kill the weed seedling is called as ________
2. Weeds are destroyed before planting of a crop is________
3. Blind tillage is usually adopted as a practice to control weeds in________
4. Preventive and eradication method are related to________
5. Hand hoeing, hand pulling, tillage and burning are related to method of________

II. Choose the best answers

1. Significant examples for bio-control of weeds are
   (a) Water hyacinth (b) Prickly pear
   (c) Lantana (d) All of these
2. Mycoherbicides
   (a) Devine (b) Biolophos
   (c) Both A & B (d) None of the above
3. Which one is mechanical weed management?
   (a) Stale seedbed (b) Hand weeding
   (c) Summer tillage (d) Soil Solarization
4. *Lantana camara*is controlled by
   (a) *Crocidosema lantana* (b) *Melanagromyza cuscutae*
   (c) *Bactra verutana* (d) *Altica cynanea*
5. Which one is method is not cultural weed control
   (a) Stale seedbed (b) Burning
   (c) Summer tillage (d) Blind tillage

III. Answer the following questions

1. Principles of weed management
2. What is stale seedbed?
3. What is soil solarization?
4. Define bio-Herbicides and give examples
5. Chemical weed management and write advantages and disadvantages.
6. Write about blind tillage
7. Cultural method of weed control

18

HERBICIDES - MERITS AND DEMERITS OF USING HERBICIDES - CLASSIFICATION BASED ON MODE OF ACTION AND METHOD OF APPLICATION - HERBICIDE FORMULATIONS - PHYTOTOXICITY

1. HERBICIDE

It also called as phyto-toxic chemical. A chemical used for killing or inhibiting the growth of targeted plants (Herbicide word derived from Latin word *herba* means plant and *caedera* means kill).

1.1. Merits of herbicides

1.1.1 On weed control

- They kill unwanted plants.
- They are easy to use
- Herbicides can be used on closely planted crops where other methods cannot be used.
- Most of the time one application of the herbicide is enough whereas other methods have to be continually used.
- They work fast. They can be removed quickly in critical situations.
- Herbicides are relatively cheap, and most of the time cheaper than manual weeding.

1.1.2 On crop growth

- They can destroy plants bearing diseases.
- They help the crops grow by destroying the weed that causes harmful effects.
- They can be safely used as manual and mechanical removing of weeds can destroy the crop.

1.1.3 Others

- They are relatively safe on lands which may erode.
- Non-selective herbicides can effectively clear fields.

1.2 Demerits of herbicides

1.2.1 Effects of herbicides on environment

Herbicides vary greatly in chemical composition and in the degree of threat they pose to the environment. .

Soil: Some herbicides are non-biodegradable and are harmful for a long period of time.

Water: The improper use of pesticides and herbicides may also cause the storm water infiltration into groundwater.

Living organisms: Herbivores may eat the plants treated with herbicides and then carnivores eat the herbivores. The toxic herbicide would be passed up the food chain increasing in concentration each time resulting in cancers and even deaths.

1.2.2 Effects of herbicides on humans

Hay fever and asthma aggravated by pollens of *Ambrosia artemissifolia. Parthenium hysterophous, Poison ivy* (*Rhus* sp), common rag weed (*Ambrosia artemissifolia)* are responsible for respiratory problems and skin allergies (dermatitis).Aquatic weeds like water lettuce (*Pistia lanceolata*), salvinia (*Salivina auriculata*), alligator weed (*Alternanthera sp*) act as alternate host and vectors of malaria, yellow fever, encephalitis, dengue fever and filariasis. Wheat flour contaminated with seeds of cocklebur gives bitter taste to the bread and irritates the gastric tract of the consumer. When the Mexican poppy seeds (*Argemene mexicana*) crushed with mustard seeds cause death and blindness of people. Milk from animals feed on the Mexican poppy weed can cause 'Glaucoma' in humans. The *Argemone* toxicity is due to an alkaloid called 'Sanguinarine' and 11-oxotriacontanoic acid.

1.2.3 Effect of herbicides on crop plant

Reduction of crop yield occur due to weeds infestation result from their multifarious ways of interfering with crop growth and crop culture. Weeds compete with crops for more plant growth factors such as mineral nutrients, water, solar energy and space and they hinder crop cultivation operations.

1. **Competition for mineral nutrients:** Being hardy and vigorous in growth habit; they soon outgrow the crops and consume large amount of water and nutrients. Thus causing heavy losses in yields. In general, weeds removed N and K from soil in much greater quantity than P.
2. **Competion for water:** For producing equal amount of dry matter, the weeds, in general transpire more water than most crop plants. It is reported that wild mustard transpires about four times more water than a crop of oat.
3. **Competition for Solar Energy:** About 99% of dry matter in plants is made up of organic matter that is dependent on solar energy. When plants are mutually shaded, their production potential is greatly reduced even though water and other nutrients are available to them in abundance.

4. **Competition for space:** Weeds compete for space both in the rhizosphere and atmosphere. In the presence of weeds, crop plants also have limited space to develop their shoots, which amounts to reduced photosynthesis in them.
5. **Weeds reduce the crop quality:** Weeds may reduce the quality of the crop produce in many ways. Weed seed like wild mustard, sweet clover, a Mexican poppy and bulblets of wild garlic and wild onion when threshed and ground with winter grains can results in serious consequences besides imparting objectionable odour to the flour. Khanna observed that striga (striga Spp.) reduced the quality of sugarcane juice by 3.9 to 8.9 percent.
6. **Weeds impair the quality of the animal produce:** Many weeds in pastures and forage crops impart undesirable flavours to milk and meat of animals. For e.g. Pivali tilwan (Cleome viscosa) imparts undesirable flavour to milk. Gokharu or Landaga (Xanthium strumarium) get attached to the body of sheep and seriously impair the quality of wool.
7. **Weeds harm animal health:** Several weeds of grasslands and forage crops contain high alkaloids, tannis, oxalates, gulcosides, and other substances that prove poisonous to animals when ingested. For e.g. Silky lupine (Lupinus sericeus) is responsible for crooked calf disease.
8. **Weeds harbour insect pests or diseases:** Weeds either give shelter to various insect pests and diseases or serve as alternate host. For e.g. Weed around paddy bunds harbour the gallfly.
9. **Weeds damage human health:** Health, comfort and work efficiency of man are also affected by weeds directly or indirectly. For e.g. people in U.P. are plagued year after year with hay fever and asthma aggravated by pollens of regweeds bursage. Tsetse fly which cause African sleeping sickness.
10. **Weeds contaminate water bodies:** Aquatic weeds change the flavour appearance and taste of drinking water. Aquatic weeds are a menace to fisheries too. Aquatic weeds on decomposition gives offensive odours and pollute atmosphere.
11. **Weeds cause quicker wear and tear to farm implements:** Being hardy and deep-rooted, the tillage implements get worn early.
12. **Weeds reduce the value of the land:** Agricultural lands heavily infested with perennial weeds like Kans (Saccharum spontaneum) always fetch less price.
13. **Less efficient use of land:** In case of perennial weeds, the carrying capacity of the grazing lands is reduced and cause depreciation of land value.
14. **Increase in cost of cultivation:** In fields of crops infested with weeds, the tillage operations require high cost.
15. **Disturbance in Public places:** It is desirable that public places be kept clean of weeds. Presence of weeds around our living and working places makes the surroundings dull.

1.2.4 Build-up of resistant biotypes

Apart from their effect on the environment, another major problem with herbicides has been the build-up of herbicide-resistant biotypes where the same herbicide has been used repeatedly for a number of years. First time, during 1968 report of weeds resistant to herbicides (triazines) was recorded. By 1991,120 weed biotypes were observed as resistant to triazine herbicides and 15 other herbicide families were documented throughout the world. Since then, nearly 500 unique cases of herbicide resistance have been reported.

2. CLASSIFICATION OF HERBICIDES

2.1. Based on method of application

1. **Soil applied herbicides:** Herbicide act through root and other underground parts of weeds. Example: Fluchloralin
2. **Foliage applied herbicides:** Herbicide primarily active on the plant foliage. Example: Glyphosate, Paraquat

2.2 Based on mode of action

1. **Selective herbicide:** A herbicide is considered as selective when in a mixed growth of plant species, it kills some species without injuring the others. Example: Atrazine
2. **Non-selective herbicide:** It destroys majority of treated vegetation. Example: Paraquat.

2.3 Based on time of application

Pre - plant application (PPI): Application of herbicide before sowing or along with sowing. Either it is foliar applied or incorporated in soil soon after its application. Pre-plant foliar spraying of glyphosate to control perennial weeds like *Cyperus rotundus* and pre-plant soil incorporation of Fluchloralin to control weeds in ground nut

Pre - emergence: Herbicide is applied to soil soon after sowing a crop before emergence of weeds. Example: Atrazine, Pendimethalin, Butachlor, Thiobencarb, Pretilachlor

Post - emergence: When herbicide is applied to kill young weeds standing in the crop plants or application after the emergence of weed and crop. Example: Glyphosate, Paraquat, 2,4-D Na Salt.

Early post emergence: Another application of herbicide in the slow growing crops like potato, sugarcane, 2-3 week after sowing is classified as early post emergence.

3. HERBICIDE FORMULATIONS

Herbicides in their natural state may be solid, liquid, volatile, non-volatile, soluble or insoluble. Hence these have to be made in forms suitable and safe for their field use. An herbicide formulation is prepared by the manufacturer by blending the active ingredient with substances like solvents, inert carriers, surfactants, stickers, stabilizers etc

3.1 Types of formulation

3.1.1. Emulsifiable concentrates (EC)

A concentrated herbicide formulation containing organic solvent and adjuvants to facilitate emulsification with water. Example: Butachlor

3.1.2 Wettable powders (WP)

A herbicide is absorbed by an inert carrier together with an added surface acting agent. The material is finely ground so that it may form a suspension when agitated with a required volume of water. Example: Atrazine

3.1.3 Granules (G)

The inert material (carrier) is given a granular shape and the herbicide (active ingredient) is mixed with sand, clay, vermiculite, finely ground plant parts (ground corn cobs) as carrier material. Example: Alachlor granules.

3.1.4 Water soluble concentrates (WSC)

Example: Paraquat

4. METHOD OF HERBICIDE APPLICATION

1. Spraying
2. Broadcasting

Factors determining the methods of application are:

- Weed-crop situation
- Type of herbicides
- Mode of action and selectivity

Depending on the target site, the herbicides are classified into

- Soil applied herbicides
- Foliage applied or foliar herbicides

Soil application	Foliar application
1. Surface	(i) Blanket spray
2. Sub surface	(ii) Directed spray
3. Band	(iii) Protected spray
4. Fumigation	(iv) Spot treatment
5. Herbigation	

Soil application

(i) **Surface application:** Soil active herbicides are applied uniformly on the surface of the soil either by spraying or by broadcasting. The applied herbicides are either left undisturbed or incorporated in to the soil. Incorporation is done to prevent the volatilization and photo-decomposition of the herbicides.

Ex. Fluchoralin – Left undisturbed under irrigated condition / Incorporated under rainfed condition

(ii) **Subsurface application:** It is the application of herbicides in a concentrated band, about 7-10 cm below the soil surface for controlling perennial weeds. For this special type of nozzles introduced below the soil under the cover of a sweep hood.

Ex. Carbamate herbicides to control *Cyperus rotundus* / Nitralin herbicides to control *Convolvulus arvensis*

(iii) **Band application:** Application to a restricted band along the crop rows leaving an untreated band in the inter-rows. Later inter-rows are cultivated to remove the weeds. Saving in cost is possible here. For example when a 30 cm wide band of a herbicide applied over a crop row that were spaced 90 cm apart, then two-third cost is saved.

(iv) **Fumigation:** Application of volatile chemicals in to confined spaces or in to the soil to produce gas that will destroy weed seeds is called fumigation. Herbicides used for fumigation are called as fumigants. These are good for killing perennial weeds and as well for eliminating weed seeds. Ex. Methyl bromide

(v) **Herbigation:** It is the application of herbicides with irrigation water both by surface, drip, sprinkler systems. In India farmers apply fluchloralin for chillies and tomato, while in western countries application of EPTC with sprinkler irrigation water is very common in Lucerne.

Foliar application

(i) **Blanket spray:** It is the uniform application of herbicides to standing crops without considering the location of the crop. Only highly selective herbicides are used here. Ex. Spraying 2,4-Ethyl Ester to rice three weeks after transplanting.

(ii) **Directed spray:** It is the application of herbicides on weeds in between rows of crops by directing the spray only on weeds avoiding the crop. This could be possible by use of protective shield or hood. For example, spraying of glyphosate in between rows of tapioca using hood to control *Cyperus rotundus*.

(iii) **Protected spray:** It is a method of applying non-selective herbicides on weeds by covering the crops which are wide spaced with polyethylene covers etc. This is expensive and laborious. However, farmers are using this technique for spraying glyphosate to control weeds in jasmine, cassava, banana.

(iv) **Spot treatment:** It is usually done on small areas having serious weed infestation to kill it and to prevent its spread. Rope wick applicator and Herbicide glove are useful here.

4.1 Herbicide application technique

Agro-chemicals *viz.,* herbicide, pesticide, fungicide, nematicide etc., used in crop production are generally applied in the form of solution or granules. At present owing to various reasons, application of all the agro-chemicals are through spray solution form.

4.1.1 Herbicide applicators

Hydraulic applicators usually need water to work. Lever operated knapsack sprayer, motorized or ground driven applicators are examples.

4.1.2 Spraying

A spray is defined as "liquid discharged into particles and scattered into dispersed droplets". Water is the most common carriers for herbicide spraying, which facilitate uniform distribution and even coverage over the area to be treated.

In a spray solution, the herbicide is dissolved, emulsified or suspended form in water. Herbicide application is directed to soil in case of pre-planting and pre-emergence herbicides and to foliage in case of post-emergence herbicides.

4.1.3 Spray volume

Total quantity of liquid applied as spray per unit area.

4.1.4 Spraying equipment

Sprayer is simply defined as "an equipment used for spraying purposes". Herbicides are generally sprayed by using hand operated sprayers. Depending upon situations power operated sprayers are also used. The basic components of the sprayer include (i) Spray tank to hold the spray solution, (ii) pump for generating the necessary pressure, (iii) power source, (iv) pressure gauge, (v) pressure regulator, (vi) agitator and (vii) delivery / discharge system

4.1.4.1 Spray tank

A sprayer has either a built in tank or separate tank to carry the spray solution. The tank should be large enough to avoid frequent refilling. Depending upon the kind of sprayer, the tank size varies from 25 to 2250 liters and sometimes even more.

A tank is provided with a big inlet at the top provided with filter and cap and water / herbicide solution poured into it through this inlet. Small tank opening can make filling and cleaning of the tank difficult. The inside bottom of the tank should be flattened or should have slope; the lowest end of the slope where the pressure pump is fitted. Then water would be collected at that point and no water /herbicide solution would be left in the spray tank as remnant.

Table 1. Details of sprayer capacity, pressure and area coverage

S. No.	Types of sprayer	Capacity (lit)	Pressure (lb /m^2)	Area covered in day of 8 hrs (ha)	Remarks
1.	Hand sprayer	0.5 - 0.75	1-2	-	Home garden
2.	Hydraulic knapsack or Backpack sprayer	12-16	10-15	0.6	Field crops
3.	Pneumatic knapsack sprayer	9-18	-	0.4	Field crops
4.	Rocker arm sprayer	-	150	0.6 - 0.8	Field crops
5.	Power sprayer	15	125	1.2-1.6	Field crops & Soil application
6.	Tractor mounted sprayer	500-1000	100	10-12	Suited for larger area
7.	Aircraft	500-750	-	200-800	For crop belt area

4.1.4.2 Pump

Any spray liquid must be atomized before it leaves the spray nozzle. A pump provides the necessary pressure for this purpose. Pressure pump is installed inside the tank, but the provision for pumping or pressurizing from outside the tank by moving the pump handle vertically up and down. The power of the single man is sufficient for developing the requisite pressure for atomization of water. Generally pressure needs to be increased four times to double the output of sprayer. The output is approximately proportional to the square-root of pressure and thus doubling the pressure, will increase output by 40 %. In case of multi-nozzle boom, pressure requirement is more and constant pumping may be highly essential. Two kinds of pumps are generally used in herbicide sprayers.

- **Air compression or Pneumatic pumps:** These pumps force air into an airtight tank containing the spray liquid. The air pressure moves the spray liquid into the nozzle where it gets atomized before leaving the sprayers as fine spray.
- **Hydraulic or Positive displacement pumps:** These pumps take in a definite volume of spray liquid and force it through the delivery system under pressure. The pumps differ in the pressure they produce.

4.1.4.3 Power source

A source of power is needed to run the sprayer pump, this source of power may be (a) manual (b) traction (c) motor or (d) tractor and aircraft engines.

4.1.4.4 Pressure regulator

Power driven sprayers are usually fitted with a pressure regulator so as to push the liquid at a constant and desired pressure. Without a pressure regulator, a nozzle will deliver more liquid at one time and less at the other in the same trip. Moreover, some herbicides are prescribed to be applied at specific pressures to obtain best results.

4.1.4.5 Agitator

Most spray tanks carry an agitator. It may be either a mechanical or hydraulic type. The agitator keeps the spray liquid components in homogenous mixture. The mechanical agitators are usually made in the shape of metal rod with a fan or rings at its distant end.

In hydraulic agitation, there is a pipe with several side holes but closed at its free-end. It is placed in the spray tank and is fed under pressure with the spray liquid from the pumps .The jet of liquid thus formed provides the necessary agitation to the spray material. Hydraulic sprayers are provided only in large, tractor-mounted spray tanks. Agitators are most essential for the application of herbicide emulsions and suspensions.

4.1.4.6 Delivery / discharge system

Water solution moves through this system from the tank to the target area under pressure and forms uniform sized droplets or mist. It consists of *(a). Delivery tube / hose, (b). Spray lance (or) spray boom with cut-off device, (c). Strainer or filter and (d). Nozzles.* In addition, delivery system has washers and gaskets at almost every joints, which make junctions air-tight to prevent leakage of water.

Delivery tube / hose

A delivery tube is usually a small plastic tube. The length (1.0 -1.5m) or just sufficient to move the spray lance around or to and fro for comfortable spraying when the tank is on the back of spray man. A long tube is unnecessary and would be a burden to the spray man to carry it all over the field since spraying by knapsack sprayer is one man's job. It is attached with the tank in one end and the spray lance on the other.

Spray lance with cut-off device

Spray lance is a metal tube and has cut-off device at its junction with the delivery tube. The lance and cut-off device are made up of the same metal *viz.*, brass, stainless steel, aluminum or copper. The cut-off when pressed by the hand allows water to flow from the delivery tube to the lance under pressure and ultimately water is discharged through nozzle. Otherwise, the flow of water remains cut-off or suspended. A lance may be fitted with normally one nozzle, but may have two or three nozzles too.

Strainer/Filter

Strainer usually filters the particles that can clog the nozzle spray tip and regulates spray flow and pattern and holds nozzle in place.

Nozzle

The nozzle is the most important part of a sprayer. The function of spray nozzle is to break the pressurized spray liquid into droplets for application to the target. Nozzles are identified by their (1) droplet size, (2) delivery and (3) spray pattern that they produce. Of the three characteristics, the spray pattern is fixed for a given nozzle (except triple action nozzle).

There are at least six major types of nozzles in use for the application of herbicides. These are 1. Flat-fan nozzle, 2. Solid cone nozzle, 3. Flood-jet nozzle, 4. Low volume nozzle, 5. Centrifugal nozzle and 6. Blast nozzle.

Table 2. Major types of nozzles and their suitability for spraying

S. No.	Nozzle type	Droplet size	Suitability	Recommended for use
1.	Flat - fan/ Deflector	Coarse	Tapered edge: soil surface application Rectangular: Band application. Controlling weeds in crop and non-crop situations	Pre-emergence and pre- planting herbicide application in soil and post emergence application of translocated type of herbicides
2.	Solid cone / Hollow	Fine	For broadcast and row crops spraying ensuring reliable coverage of the whole plant	Insecticide and fungicide, spot application, post emergence application of contact type herbicides
3.	Flood jet	Medium	Ideal for uneven surface and windy condition, reducing drift hazard of herbicide	Surface as well as sub-surface application of herbicides
4.	Low volume nozzles	Coarse	Wide spaced crop as well as fallow land	Herbicide without much dilution
5.	Centrifugal nozzle	Fine	Spraying viscous liquids like invert emulsions	Emulsifiable, high concentrated and high cost herbicides
6.	Blast nozzle	Fine	Closed spaced as well as tall growing crops	Insecticide and fungicide, pre and post emergence application through motorized mist blower

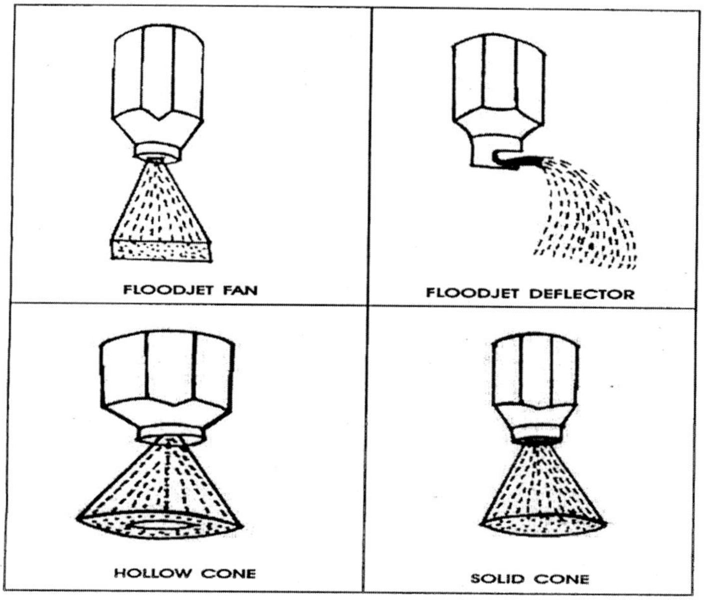

Fig. 13. Different types of nozzles for herbicide application

5. CALIBRATION OF SPRAYER

The process to find out the exact quantity of carrier (water) required for spraying a given area under a given set of conditions is called calibration. It is done to apply the correct amount of herbicide over a given area under a given set of condition. The best way to calibrate is to spray an area of known size and measure the volume of spray delivered from the tank.

(i) Quantity of spray fluid required $(Q) = \dfrac{(V_1 - V_2)}{a} \times A$

Where,

V_1 = Volume of water in the sprayer before spray (taken initially)

V_2 = Volume of water in the sprayer after spray (remained after spraying)

a = Measured area of spray (m^2)

A = Area to be sprayed (m^2)

(ii) The area covered per hour can be calculated as shown under:

$$\text{Area coverage (ha hr}^{-1}) = \frac{\text{Walking speed (m/hr)}}{10000} \times \text{Spray width (m)}$$

If a person walks 1 km hr^{-1} covering a swath of 0.6m, the area covered 0.06 ha hr^{-1}. At this rate, it will take 16 hr 40 min. to cover one hectare. At a spray discharge rate of 30 lit hr^{-1} it would require a spray volume of 500 l ha^{1}.

(iii) Pump capacity can be calculated by the following formula:

$$\text{Pump capacity} = \frac{\text{Spray volume (lit/ha)} \times \text{Swath width (metre)} \times \text{Walking speed (km/hr)} \times \text{Metre/kilometre}}{\text{Metre square/hectare} \times \text{Minute/hour}}$$

6. STUDY ON PHYTOTOXICITY SYMPTOMS OF HERBICIDES IN DIFFERENT CROPS

Proper diagnosis of herbicidal injury to plants particularly to crops is a difficult task. Diagnosis becomes complex and complicated since the symptoms of herbicidal injury may resemble to the symptoms due to many stress factors such as infectious diseases, air pollution, other pesticides, nutrient imbalances (deficiency or toxicity), adverse environmental conditions like moisture stress and frost injury. The merger of these symptoms together is also quite possible under field conditions, which leads to more confusion regarding identification of the correct symptoms of herbicidal injury. The visual symptoms of herbicidal injury are caused by the deficiency of certain metabolites or substances required for the growth of plants. They are described below:

(i) **Wilting:** Wilting and desiccation of plants are common symptoms of membrane disruption resulting from diffusion or leakage of solutes from cells to the intercellular spaces followed by plasmolysis. Metsulfuron - methyl, 2, 4-D, imazethapyr, tribenuron-methyl induces wilting in plants.

(ii) **Chlorosis followed by necrosis:** Leaf chlorosis followed by necrosis is a common response of plants on exposure to the photosynthesis - inhibitor herbicides. Almost more than 50% of the herbicides used in agriculture are photosynthesis inhibitor directly or indirectly. They include herbicide groups such as, phenylureas, uracils, triazines and herbicides like propanil, pyrazon, bentazon, sulfosulfuron.

(iii) **Necrosis:** Here only necrosis occurs and no chlorosis. Paraquat, diquat, bromoxynil, ioxynil, nitrofen, endothall, fluorodifen, cacodylic acid and acrolein induce necrosis (Table 3). They all are contact herbicides and cause disruption of cellular membrane in plants.

Table 3. Major morphological changes induced by various herbicides

Leaf chlorosis[a]	Necrosis[b]	Epinasty[c]	Formative effects[c]
Triazines[d]	Acrolein	Phenoxyalkanoic acids	Acetamides, Dinitroanilines
Uracils[d]	Bromoxynil	Dicamba	Nitriles, Thiocarbamates
Phenylureas[d]	Cacodylic acid	Fenac	Barban, Bensulide, Chlorpropham
Dalapon[e]	Dinoseb	Picloram	Dalapon, DCPA, Diphenamid
Desmedipham[d]	Diquat	2,3,6-TBA	Glyphosate, Naptalam
DSMA[e]	Endothall		Phenoxyalkanoic acids
Phenmedipham[d]	Fluorodifen		Propham, Pronamide, TCA

Propanil[a]	loxynil		
Pyrazon[a]	Nitrofen		
Ghyphosate[e]	Paraquat		
MSMA[e]	TCAC		

Where,

- a - may be followed by necrosis,
- b - usually associated with rapid membrane damage without chlorosis
- c - associated with altered cell division, cell enlargement and tissue differentiation
- d - chloroplast and membrane damage
- e - probably associated with senescence

(iv) Leaf burning: Paraquat and diquat cause burning of leaves by rapid desiccation. Even isoproturon produces this symptom on *Avena ludoviciana* (wild oat) and *Phalaris minor* (little seed canary grass). The leaves start burning from the top/ apex.

(v) Stunted or retarded growth: Growth of plants becomes very much stunted as a result of inhibition of cell elongation. Chlorsulfuron, benzoylprop-ethyl, flamprop-isopropyl, flamprop-methyl and sulfosulfuron causes retarded growth of grass weeds, namely *Avena ludoviciana* and *Phalaris minor* in winter cereals mainly wheat.

(vi) Epinasty / epinastic symptoms: Epinasty means bending, twisting or curling of leaves, stems or even ear head, which later become brittle and break up even by light pressure. Epinastic effects are usually associated with altered cell division, cell elongation and tissue differentiation (Table 3). Ex. 2, 4-D, MCPA, MCPB (Phenoxyl alkanoic acids), 2,3,6-TBA, dicamba (Benzoic acids), fenac (Phenyl alkanic acids) and picloram Picolinic or Pyridinecarboxylic acids). They may also cause adventitious roots, callus or fascinated organ formation.

(vii) Formative effects: Formative effects like epinastic effects are also associated with the altered cell division, cell elongation and tissue differentiation (Table 3). Chloro-acetamides, dinitroanitines, nitrites, thiocarbamates, barban, bensulide, chlorpropham, dalapon, DCPA, diphenamid, glyphosate, naptalam, phenoxy alkanoic acids, propham, pronamide, TCA produces formative effects in plants.

- Increased tillering due to apical portion damaged
- Multiple shoots formation at one node
- Shorter internode, thickened leaves, coleoptile and leaf distortion
- Abnormal seedling elongation
- Reduced secondary root formation
- Albinism of leaf occurs following treatment with amitrole, flouridone, norfiurazone, clomazone (in wheat) on newly developing leaf choloroplast formation is affected

- Ear head malformation: 2, 4-D causes ear head malformation in some varieties of wheat Ex. Arjun (HD 2009), Kundan. Percentage of occurrence of malformed ear head is very less. The malformed ear head is not upright or straight, usually bent and twisted having a bit disrupted look.

Crops

All the recommended herbicides are safe for respective crops in which they are recommended provided they are used according to the recommended technology. Otherwise, crops may show toxicity symptoms under certain situation such as

- **Wrong time of application:** Application of herbicides too early than its recommended time may cause yellowing and necrosis of crop. Over dose of 2, 4-D and its application on very young crop may cause tubular leaves and deformed wheat ears.
- **Use of over dose:** Application of higher than recommended dose is also responsible for stunting, yellowing and necrotic growth of crop plants
- Wrong application technology: Uneven application of herbicides not only harms the present crop but may also prove harmful for the succeeding crops due to residual toxicity.
- **Spray drift:** Due to speedy winds on the day of spray, adjoining crops may be damaged.
- **Adverse climatic conditions:** Due to heavy rains immediately after the application of herbicides, few root uptake herbicides may even damage the tolerant crops.
- **Wrong herbicide used:** With the use of unapproved herbicides, crop is damaged. For instance, wheat crop may be damaged with the use of metribuzin. Germination of wheat crop is adversely affected when seeds come in contact with triallate treated soil zone.

Weeds

Toxicity is adverse effect of herbicide on the growth and development of plants.

Type of Toxicity

It is of two types

1. **Acute Toxicity:** When the effect of herbicide is not having quick effect and shows its phytotoxic treatment. Ex. Effect of gramoxone
2. **Cronic toxicity:** When a herbicide is not having quick effect and shows its phytotoxic effect after few days of application. EX. Effect of isoproturon on *Phalaris minor*.

QUESTIONS

I. Fill in the blanks

1. ________ used for killing or inhibiting the growth of targeted plants.
2. Chemical in commercial product that is directly responsible for the herbicides activity is called as ________
3. Kills all vegetative when comes in contact is known as ________
4. Application of herbicides after planting is known as ________
5. Application of herbicide before sowing or along with sowing is known as ________

II. Choose the best answers

1. Which one is foliage applied herbicides?
   (a) Fluchloralin (b) pendimethalin
   (c) Glyphosate (d) Paraquat
2. Which herbicides are come under pre plant application?
   (a) Fluchloralin (b) Alachlor
   (c) Both a & b (d) Atrazine
3. Which herbicides are come under post emergence application?
   (a) 2, 4 –D (b) Paraquat
   (c) Propanil (d) All the above
4. Pendimethalin is a ________herbicide
   (a) Selective (b) Non Selective
   (c) Postemergencege (d) *Foliage*
5. Which is a selective herbicide?
   (a) Pendimethalin (b) Butachlor
   (c) Atrazine (d) All the above

III. Answer the following questions

1. Define herbicide and write merits and demerits
2. Write classification of herbicides
3. What is herbicide formulation? and write their types
4. Differentiate pre and post emergence herbicide with examples
5. Write the method of herbicide application

19

WEED MANAGEMENT PRACTICES FOR MAJOR FIELD CROPS

1. CEREALS AND MILLETS

1.1 Rice

1.1.1 Yield losses

Reduction in yield to the tune of 34% in transplanted rice, 45% in direct seeded low land rice and 67% in upland rice are reported.

1.1.2 Critical period for weed

The critical period for weed free condition for higher productivity is reported to be 30 – 35 days in transplanted rice where as direct seeded low land and upland condition the weed free period ranges from 40-60 days.

1.2 Upland rice

1.2.1 Pre emergence herbicides

- Cyhalofopbutyl 0.06-0.07(G) and Metsulfuron 0.03-0.06kg/ha (BLW).
- Where as oxadiazon @ 0.6 to1.5, butachlor 1-1.5, oxadiargyl 0.09 kg/ha at 7-8 days, quinclorac 0.187-0.375, dithiopyr (crabgrass)@ 0.18 kg/ha to control grasses and broad leaved weeds.
- Pretilachlor 0.75kg/ha control grasses. broad leaved weeds and sedges also.

1.2.2 Post emergence herbicides

- Propanil (3-4 kg a i/ha) effective at 2-3 leaves stage of G + BLW
- 2,4-D@ 0.5 -1.0 kg a.i /ha effective to control established broad leaved weeds. In addition to this for broad leaved weed control ethoxysulfuron (sunrise) @ 0.03 kg a.i/ha (75 gm /ha) applied at 20 days.
- Chlorimuronethyl and metsulfuron @ 0.004 and 0.03-0.06 kg/ha respectively to control broad leaved weeds.
- Fenoxaprop-p-ethyl (puma super) 0.05-0.075kg/ha is effective to control both grasses and sedges.
- Broad spectrum weed control can be achieved by pretilachlor 0.75 and pyrazosulfuron ethyl 0.025kg/ha to control Broad leaved weeds and sedges.

1.3 Low land /transplanted rice

1.3.1 Weed management in nursery

- Hand pulling is the common method of weed control in nurseries.
- Application of butachlor (0.75-1.0kg/ha) or thiobencarb (1.5-3.0kg/ha) or pretilachlor + safenor (sofit) (0.75kg/ha), anilofos (0.25-0.5kg/ha) at 4-7 days before or after sowing through ponded water where as oxadiazon (0.5-0.75kg/ ha)or cyhalofop butyl (clincher)0.1 kg a.i/ha on 8th day after sowing results in effective control of weeds
- Nominee gold@0.002 kg a.i/ha (200 ml/ha) at 14 – 15 daysto control of weeds.

1.3.2 Weed management in main field

Pre emergence application

- Butachlor @ 1.5-2 kg/ha, anilophos 0.25-1.0 kg/ha, oxadiargyl 0.07-0.125 kg/ ha, clomazone 0.2-0.25 kg/ha and acetachlor 0.10-0.15 kg/ha to control grasses and broad leaved weeds.
- Whereas 2,4-D easter G (0.8-1.2kg/ha) controls sedges and broad leaved weeds and chlorimuronethyl (0.01kg/ha) to control broad leaved weeds. Herbicides should be applied 4-7DAT. The field should not be drained till 7 days to obtain satisfactory weed control.

Post emergence herbicides

- Post emergence herbicides are effective at 4-6weeks after transplanting. The field should not be drained before the application of post emergence herbicide. Fenoxaprop-P-ethyl 0.075-0.120, cyhalofop butyl 0.06-0.75 kg/ha to control grasses and ethoxysulfuron 0.02kg/ha for BLW should be applied 10DAT.
- Cinosulfuron (0.02kg/ha) used as pre and post emergence herbicide to control grasses and broad leaved weeds. Where as 2,4-D and MCPA@ (0.75-1.5kg/ha) used to control sedges and broad leaved weeds.
- For controlling unwanted algal growth in the rice fields copper sulphate and copper oxy chloride @ 8-10kg/ha have been used since long time. But now more effective products like brestan -60 (0.07-1.7 kg a.i /ha) and potassium azide are available for this purpose.

1.2 Wheat

The weeds reduce grain yield up to 10-40% and competition is during first 30-40 days after sowing the crop.Hand weeding twice at 15 days interval because of narrow row spacing is recommended Several herbicides are used in conjunction with good crop husbandry to control specific weeds.

Pre emergence: Diuron and linuron @ 0.5 -1.0 kg kg/ha are effective to control grasses, broad leaved weeds, where as linuron is effective to control sedges also. Pendimethalin and trifluralin @1.0 &0.5 -1.0 kg/ha respectively to control grasses, broad leaved weeds.

Dicots: 2,4 D (EE) 0.3-0.4 kg /ha at 35DAS;

Monocots: Isoproturon 1-1.5kg / Methabenzthiazuron 1.5 kg / Metoxuron 1.5 kg/ ha 30-35 DAS

Pre-emergence application of Pendimethalin or Isoproturon is broad spectrum control

Alternate herbicides for control of isoproturon resistance *Phalaris minor*

Metribuzin @ 0.245-0.315kg/ha as pre & post and sulfosulfuron @0.025-0.05kg/ha herbicides are effective for both grass and broad leaved weeds. Whereas fenoxaprop-P- ethyl (0.062-0.12kg/ha) and mesosulfuron (0.01-0.030kg/ha), and tralkoxydim (0.35-0.4kg/ha) control grassess like *Phalaris minor* and *Avena fatua* than broad leaved weeds.

1.3 Maize

- Crop weed competition is 6 weeks
- Maize yield was reduced as much as 25-80%.
- Ridges and furrow method and subsequent earth up can manage weeds effectively
- Under rainfed flat system two hoeings and earth up are essential
- Pre-plant incorporation of butylate or EPTC @ 3-4 kg/ha mixed with 0.5 kg/ha of atrazine or simazine controls nut grass and many annual grassess.
- Pre-emergence application of atrazine and simazine @ 1-2 kg/ha. to control grasses and broad leaved weeds

1.4 Sorghum

- Weed free situation for 5 weeks
- Herbicides
- Atrazine 0.25 kg /ha for pure crop
- Pendimethalin for pulses inter cropped situation

1.4.1 Striga management

(Witch weed) (*Striga asiatica*, *S. lutea*, *S. hermonthica*)

Control measures

- Post emergence application of 2,4 D Na salt 2.0kg/ha at 25-30 DAS
- Intercropping with redgram
- Crop rotation with trap crop likes sunflower, groundnut, cowpea – they induce weeds to germinate but not affected

- Heavy application of N and FYM
- Spraying urea 10% solution 25-30 DAS
- Using germination stimulants like strigol andethylene

1.5 Ragi

- Apply PE Oxyfluorfen @ 0.05 kg a.i/ha on 3 DAS using backpack knapsack/ rocker sprayer fitted with flat fan type of nozzle with 500 litre of water/ha followed by one hand weeding on 20 DAS.
- Apply the herbicides when there is sufficient moisture in the soil or irrigate immediately after application of herbicide.
- If pre-emergence herbicide is not applied, hand weed twice on 10thand 20thday after transplanting.
- For rainfed direct seeded crop, apply post emergence herbicide; 2,4-DEE or 2,4-D Na salt at 0.5 kg/ha on 10thday after sowing depending on the moisture availability.

1.6 Minor millets

First weeding on 15th DAS and the second on 40th DAS (or) apply Isoproturan @ 0.5 kg on 3rd DAS and 2,4 D Na salt @0.5 kg on 20-25 DAS.

2. PULSES

2.1 Blackgram and greengram

- Pre emergence application of Pendimethalin 3.3 litres/ha under irrigated condition or 2.5 litres/ha under rainfed condition on 3 days after sowing using Backpack/ Knapsack/Rocker sprayer fitted with flat fan nozzle using 500 litres of water for spraying one ha followed by one hand weeding at 20 DAS (or) EPOE application of quizalofop ethyl @ 50 g ai/ha^{-1} and imazethapyr @ 50 g ai ha^{-1} on 15 – 20 DAS. If herbicides are not applied give two hand weedings on 15 and 30 days after sowing.
- For the irrigated blackgram PE isoproturon @ 0.5 kg ha^{-1} followed by one hand weeding on 30 DAS.

2.2 Redgram

- Pre emergence application of Pendimethalin 0.75 kg/ha (2.5 litres/ha) on 3 DAS mixed with 500 litres of water using backpack/knapsack/rocker sprayer using flat fan deflector type of nozzle. Then irrigate the field. Following this, one hand weeding may be given on 30-35 DAS
- If herbicide is not given, give two hand weedings on 20 and 35 DAS.
- In case of labour problem, apply Pendimethalin 0.75 kg (2.5 lit/ha) on 3 DAS followed by early post emergence application of Imazethapyr @ 60 g ai/ha on

15 DAE of weeds (2 - 3 leaves stage of weeds) and quizalofop ethyl @ 50 g ai/ha on 20 DAE of weeds (2 - 3 leaves of weeds) are recommended for controlling broad leaved and grassy weeds, respectively. If both the weeds are present, tank mix application of Imazethapyr @ 60 g ai/ha and quizalofop ethyl @ 50 g ai/ha at 15 - 20 days after emergence of weeds (2 - 3 leaves stage of weeds) is recommended.

- Apply PE metalachlor 1.0 kg ha^{-1} on 3 DAS followed by one hand weeding on 40 DAS. Note: At the time of herbicide application, there should be sufficient soil moisture.

2.3 Soybean

- Alachlor may be applied to the irrigated crop at 4 litres/ha or Pendimethalin 3.3 litre ai/ha after sowing followed by one hand weeding on 30 days after sowing.
- If herbicide spray is not given two hand weedings on 20 and 35 days after sowing may be given
- Imazythypur may be applied as post emergence on 20 DAS with one hand weeding on 30 days after sowing.

3. OILSEEDS

3.1 Groundnut

Critical period for weed growth is 20-45 DAS. Losses as high as 70%

1. Below 40 days 1-2 hand weedings followed by intercultivation for bunch type of groundnut
2. Pre plant incorporation of fluchloralin @ 1-2 kg/ha to control broad leave weeds.
3. Pre-emergence application of Pendimethalin (2 kg/ha) or metolochlor (0.75-1.0kg/ha) or butachlor (1.0kg/ha) or nitrofen (2-4 kg/ha) or oxadiazon (1-2kg/ha)
4. Post-emergence application of Fluazifop (0.125 – 0.250 kg/ha)30-40 days after sowing groudnut against grasses especially *Cynodon dactylon*
5. Imazethpyr @0.75kg/ha for control of mixed growth of grasses and broad leaved weeds.
6. Quizalofopethyl 0.4-0.5kg/hafor control of annual perennial grass weeds.

3.2 Gingelly

Sesame is sensitive to weed competition during the first 15-35 DAS. A minimum of two weedings, one after 15 DAS and another 35 DAS are required to keep the field relatively weed free. Row seeded crop facilitates use of blade harrows for intercultivation. Two intercultivations, 15 and 35 DAS followed by one hand weeding keeps the field free of weeds.

Herbicides use, especially under rainfed conditions, is very limited due to low yield, which may not compensate for the cost of herbicides. If necessary, Alachlor (1.0) or Thiobencarb (2.0) can be used as PRE spray for effective control of weeds. Use of PRE herbicides followed by on hand weeding around 30 DAS is the most appropriate way of weed management in sesame.

3.3 Sunflower

1. Can be grown in all seasons
2. It is grown in rice fallows
3. Critical period is 4-6 weeks after sowing
4. Pre-plant incorporation of trifluralin @1kg/ha to control grasses and broad leaved weeds or fluchloralin @ 1.0 1kg/ha pre-plant incorporation to control broad leaved weeds.
5. **Pre-emergence application:** Pendimethalin (0.75-1.0kg/ha kg/ha) or Metolachlor 1.0 kg/ha kg/ha or Alachlor 1.0-2.0 G or Oxyfluorfen 0.25kg/ha or Butachlor 1-1.5kg/ha
6. **Post-emergence application:** Fluazifop-P butyl @ 0.25 kg/ha 21- 25DAS to control grasses.

4. COMMERCIAL CROPS

4.1 Cotton

4.1.1 Cotton (irrigated)

- Apply Pendimethalin @ 3.3 l/ha three days after sowing, using a hand operated sprayer fitted with deflecting or fan type nozzle. Sufficient moisture should be present in the soil at the time of herbicide application. This will ensure weed free condition upto 40 days.
- One hand weeding on 45 DAS will keep weed free environment upto 60 DAS.
- Hoe and hand weed between $18^{th}$to 20^{th}day of sowing, if herbicide is not applied at the time of sowing.

4.1.2 Cotton (Rainfed)

- Pre-emergence application of Pendimethalin 3.3 l/ha followed by one hand weeding on 40 days after crop emergence. At the time of herbicide application sufficient soil moisture must be there.
- If sufficient soil moisture is not available for applying herbicides hand weeding may be given at 10 - 20 days after crop emergence.

4.1.3 Rice fallow cotton

- Pre-emergence application of Pendimethalin 3.3 l/ha ensures weed free condition for 40 - 45 days. This should be followed by one hand weeding and earthing up during 40 - 45 days.
- Take up this operation when the top soil dries up and comes to proper condition.

4.2 Sugarcane

- Wherever weed menace is higher, one line weeding along the crop row and spade digging of ridges have to be done on 30, 60 and 90 DAP
- Spray Atrazine 2 kg or Oxyflurofen 750 ml/ha mixed in 600 liters of water as pre emergence herbicide on the 3rd day of planting, using deflector or fan type nozzle fitted with knapsack sprayer.
- The pre emergence application of atrazine @ 1.0 kg a.i. ha^{-1} on 3 DAP followed by post emergence directed application of glyphosate @ 1.0 lit ha^{-1}on 45 DAP with hood+ one hand weeding on 90 DAP registered the maximum cane yield.
- If pre-emergence spray is not carried out, post-emergence spray of Grammaxone litre + 2,4-D sodium salt 2.5 kg/ha in 600 liter of water on 21st day of planting.
- If the parasitic weed *striga* is a problem, post-emergence application of 2,4-D sodium salt may be done. 2, 4-D spraying should be avoided when neighboring crop is cotton or bhendi. Apply 20% urea also for the control of *striga* as direct spray.
- Pre- plant application of glyphosate at 2.0 kg ha^{-1} along with 2% ammonium sulphate at 21 days before planting of sugarcane followed by post emergence direct spraying of glyphosate at 2.0 kg ha^{-1} along with 2% ammonium sulphate with a special hood on 30 DAP suppressed the nut sedges *(Cyperus rotandus)* and provided weed free environment.
- If herbicide is not applied work the juniorhoe along the ridges on 25, 55 and 85 days after planting for removal of weeds and proper stirring. Remove the weeds along the furrows with hand hoe. Otherwise operate power tiller fitted with tynes for intercultivation.
- Control of creeper weeds post emergence directed application of fernoxone (2, 4 –D sodium salt) @ 2 gm + 10 gm of urea per liter of water may be sprayed over the creeper weeds.

4.2.1 Weed management in sugarcane intercropping system

Premergence application of Thiobencarb @ 1.25 kg ai/ha under intercropping system in sugarcane with soybean, blackgram or groundnut gives effective weed control.

QUESTIONS

I. Fill in the blanks

1. Critical period for weed free condition in rice is_____
2. Average lose due to weeds for different crops is_____
3. The critical period for crop – weed completion in most of the crops is ________
4. Herbicide recommended for cotton________
5. Post emergence of herbicide applied in _______stages of weeds

II. Choose the best answers

1. *Echinocloa colonum* and *E. Crusgali* both is associated weed with
   (a) Rice (b) Maize (c) Barley (d) Wheat
2. Which herbicides used to control weeds in maize?
   (a) Simazine (b) Atrazine (c) Both A &B (d) Alachlor
3. Post emergence herbicide used for rain fed ragi is
   (a) 2,4-D Na @ 0.5 kg (b) 2,4-DEE @ 0.5 kg
   (c) Both A& B (d) Isoproturan@ 0.5 kg
4. Pendimethalin recommend for ___________
   (a) Cotton (b) Sunflower (c) Pulses (d) All the above
5. Critical period for weed free condition for higher productivity of transplanted rice is
   (a) 20- 30 DAS (b) 30-35 DAS (c) 25-30 DAS (d) 20 -25 DAS

III. Answer the following questions

1. Integrated weed management in rice
2. Weed management in rice nursery
3. Integrated weed management in pulses
4. Integrated weed management in cotton
5. Integrated weed management in sugarcane

20

WEED MANAGEMENT PRACTICES FOR PARASITIC WEEDS, PROBLEMATIC WEEDS AND AQUATIC WEEDS

1. PARASITIC WEEDS

1.1 *Striga asiatica* (Sudu malli)

Semi root parasite in cereal crops like sorghum, maize, varagu and sugarcane. Sucks water and nutrients from the hosts directly from the vascular system. Striga seeds are very minute, 1/20th size of tobacco seeds. Single plant may produce 1500 seeds.

Biology and habitat

It is much more tropical in their distribution. Smallness of their seeds make them obligate parasite only able to develop fully after attachment to a suitable host root. The seeds then germinate, attached to tree roots of host plant and develop as total parasite below ground but when light is available, chlorophyll is formed and plants can photosynthesize and no longer parasitic. The efficiency of photosynthesis is only about 20 percent of that of a normal plant.

There are 23 species of *Striga* on record of which only 3 are considered notorious. They are S.*asiatica, S.hermonthica* and *S. densiflora*, attacking sorghum and bajra and all are found in Africa. *S. asiatica* is common in Asia only. In Maharastra, Karnataka and Gujarat, Sorghum is severly attacked by the *Striga* resulting in a heavy reduction in yield. Heavy losses also occur in Tamil Nadu, Andra Pradesh and certain parts of Mathya Pradesh.

Control measures

- Stimulating seed germination and destroy by tillage or herbicide
- Proper crop rotation with trap crops and catch crops
- Trap crops like soybean, cotton, cowpea, ground nut, chickpea, pigeon pea etc. are used in rotation with catch crops viz. sorghum, maize and millets and trap crop induce germination by producing stimulants but not parasitized. Catch crop induces germination and get parasitized.
- Pre-emergence application of atrazine 1.0 kg/ha on 3rd DAP + hand weeding on 45 DAP with an earthing up on 60 DAP combined with post-emergence spraying of 2,4-D at 6 g (0.6%) + urea 20 g (2%) / litre of water on 90 DAP + trash mulching 5 t/ha on 120 DAP.

1.2 *Orabanche* (Pukaielai kalan)

Orobanche, commonly known as broomrape, is a genus of over 200 species of small parasitic herbs, mostly native to the temperate Northern Hemisphere. It is the type genus of the broomrape family, Orobanchaceae.

It is a total root parasite infesting tobacco and a number of solanaceous plants.

Biology and habitat

Orobanche is originated in the near east and it is an annual parasitic herb propagating by seed. The host roots exudates and induce germination of broomrape seeds within the soil and the adult plant produces vast quantities of minute seeds (0.25 mm long). A haustorium is formed which penetrates to a vascular tissue and establishes connections with the host xylem and no chlorophyll is formed and the parasite is totally dependent on the host for its nutrition and there is interference with the hosts ability to take up water.

These are four major parasitic species of broomrape. These are; O. cernua on tobacco and sunflower, O.ramossa and O.aegyptiaca on tomato and like crops.Broomrape is an annual, parasitic herb propagating by seeds. The host root exudates induce germination of broomrape seeds within soil. The parasite seedling then infects the nearby host roots forming haustoria on them. Soon thereafter the broomrape emerges through the soil as pale shoots, devoid of chlorophyll. Broom rape is thus a total parasite. Its clones are fleshy at the base, and they develop several tubercles which grow into separate flowering shoots. Each broomrape plants may produce over half a million seeds in a short period of about eight weeks. The broomrape seeds disseminate by wind, birds and farm animals. They can remain dormant in soil for 2-12 years.

Control measures

- A long rotation with tobacco crop once in three years preceded by a chilli crop can reduce the loss due to the parasite.
- Pre-emergence application of pendimethalin 1.0 kg/ha or oxyfluorfen 0.30 kg/ha on 3 DATP in tobacco, tomato and brinjal and 3 DAP in potato.
- Directed application of paraquat 6 ml/litre of water or glyphosate 8 ml/ litre of water or imazethapyr 3 ml/ litre of water on the *Orbanche* shoots
- Plant hole application of neem cake 25 g / plant or drenching of copper sulphate 5% provides partial control of *Orabanche* in tobacco.

1.3 Loranthus

It is a semi-stem parasite on trees like citrus, mango, teak, casuarinas etc.

Biology and habitat

Loranthus is a stem parasite of certain tropical and sub- tropical trees and bushes like the teak, rosewood,sandlewood,mango,citrus and tea. In India about 60 species of *Loranthus* are on record of which *L. longiflorus var. Falcatus is* the most damaging. *Loranthus* seeds are spread by birds to fresh tree branches.

Control measures

Combination of Copper sulphate 7 g and 2,4-D 1.0 g per tree applied in the holes made on the stem of trees are effective.

1.4 *Cuscuta* (Dodder)

It is a total stem parasitic annual plant with twining thread like yellow or orange stem and leaves. The plant twines the host like lucerne very closely.

Biology and habitat

The seeds of cuscuta germinate spontaneously without involvement of any host plant and once attached to its host, the cuscuta seedlings grows very rapidly. If no host is contracted the seedling will lose capacity to parasitize. A single dodder seedling can grow 2 km or more in length.

There are at least 170 species of cuscuta in the world. The main agriculturally important species are *Cuscuta compestris, C. reflexa, C.hyalina, C. monogyna, C.chinensis, C.planiflora, C. pproximate, C. pedicillata, C. palaestina, C.epithymum, C.epillinum. Dodder* dissiminates by seeds as well as through movement of its fragments.

Control measures

- Planting crops other than legumes and flax for several years to reduce weed seed in the soil,
- Application of CIPC (chlorpropham) at 6 to 8 kg/ha in lucerne crop infested with Cuscuta
- Application of paraquat at 1.0 kg/ha over infested lucerne
- Pronamide application at 1.5 kg/ha 15 days after Cuscuta germination.

2. PROBLEMATIC WEEDS

2.1 *Cynodon dactylon* (Arugu) and 2.2.*Cyperus rotundus* (Koarai)

Biology and habitat

Cynodon dactylon is a perennial weed growing largely from rootstocks and stolons. Although it produces seeds, yet these are not important in its dispersal. Bermuda grass rootstocks from dense sod inside the soil and its stolons creep over the land, extensively. The grass grows round the year, but it is particularly vigorous under warm and moist conditions.

Management of perennial weeds like *Cynodon dactylon* by the application of Glyphosate 10 ml + AGF activator 2 ml / lit of water (or) Glyphosate 15 ml + 20 g Ammonium sulphate + activator 1 ml / lit of water.

Approach	:	Post emergence, total, translocative herbicide
Stage of weed	:	Active growing, pre flowering stage
Sprayer	:	Hand operated Knapsack / backpack
Nozzle	:	WFN 24 & ULV 50 with 30 Psi
Spray volume	:	200-250 litre / ha

Application technology

Non-Crop situation — Blanket application
Cropped situation — Pre-sowing / planting
Stale seed bed — Blanket application
Established Crops — Directed application using hoods

Note: Rain free period / waiting period: 48 hour

2.2 *Parthenium hysterophorus* (Parthenium natchu chedi)

Parthenium hysterophorus is a much-branched, short-lived (annual), upright (erect) herbaceous plant that forms a basal rosette of leaves during the early stage of growth. It usually grows 0.5-1.5 m tall, but can occasionally reach up to 2 m or more in height.

Biology and habitat

Parthenium hysterolphorus is an annual invasive alien plant, with wide amplitude of ecological adoptability; being both photo and thermo-insensitve.It reproduces it self freely from numerous seeds (5000-10000 per plant), beside its crown bud which put forth new shoots as and when the mother shoot is cut. Thus, the weed spreads by leaps and bounds in any new, neglected area. The weed is notorious for causing allergic dermatitis and mental depressions in human beings. the main toxin responsible for such effects of the weed is parthenin, present in a concentration as high as 0.33 % (d.w.b). In dry summer months Parthenium tends to stunt its growth and remain in rosette form. It shoots up new growth in the rainy season and grows fast through the winter, attaining a height about 90 cm, with profuse branching. The plant flowers and sets seeds throughout its growing period.

Control measures

- Manual removal and destruction of *Parthenium* plants before flowering using hand glouse / machineries (or)

- Pre-emergence application of atrazine 4 g / litre in 500 litres of water / hectare (or)
- Uniform spraying of sodium chloride 200g + 2 ml soap oil / litre of water (or)
- Spraying of 2,4-D sodium salt 8 g or glyphosate 10 ml + 20g ammonium sulphate + 2 ml soap solution / litre of water before flowering (or)
- Post-emergence application of metribuzin 3 g / litre of water under non crop situation.
- Raising competitive plants like *Cassia serecea* and *Abutilon indicum* on fallow lands to replace *Parthenium* (or)
- Biological control by Mexican beetle which is very active during only in monsoon seasons

Note: *Parthenium* can be decomposed well before flowering and used as organic manure.

2.3 *Solanum elaegnifolium* (Kattu kandan kathiri)

Biology and habitat

Silver leaf nightshade is a deep rooted, perennial weed with erect and branched stems covered with fine hair and orange colour prickles. Its leaves are silvery white due to numerous white hairs. The flower is purple to violet. The plant grows profusely during the rainy season, infesting roadsides, garden lands and neglected dry lands. Its root pieces can give rise to new plants In India, the weed is presently noxious in her southern states .The weed is native is South West USA and Mexico. It is poisonous to livestock.

Control measures

Post-emergence application of Glyphosate 20 ml + Ammonium sulphate 20g/ha+ activator 1ml/litre of water or Glyphosate 10 ml in combination with 2,4-D sodium salt 6 g + activator 1ml / litre.

Note: The application should be during the active growth / vegetative phase of weed

3. AQUATIC WEEDS

Weed plants growing either in water or having very high requirements for moisture are termed as aquatic weeds.

3.1 *Eichornea crassipes* (Water hyacinth or Agaya thamarai)

It is a perennial, free floating weed. The seeds germinate on the banks of water bodies and the seedlings then move to the water by water waves. Best time of application is summer.

- Manual / Mechanical removal and drying
- Application of 2,4-D sodium salt at 8g + urea at 20g or Paraquat at 6 ml / litre of water
- Application of Glyphosate 15ml+Ammonium sulphate 20g/ litre of water

Note: Vermi-composting and composting of dried water hyacinth and can be used as organic manure in irrigated upland ecosystems.

QUESTIONS

I. Fill in the blanks

1. __________ is semi root parasite host in sorghum, maize, varagu and sugarcane
2. Total root parasite infesting tobacco is __________
3. Semi-stem parasite host on trees like citrus, mango, teak, casuarinas is ________
4. Biological control of parthenium by ___________
5. Botanical name of water hyacinth ____________

II. Choose the best answers

1. Which crop associated with *striga*?
   (a) Sorghum (b) Sugarcane (c) Maize (d) Wheat
2. Which is a absolute weeds?
   (a) *Cyprus rotandus* (b) *Cyandon dactylon*
   (c) *Acalypha indica* (d) *Parthenium*
3. Which chemical recommended for *Parthenium*control?
   (a) Atrazine 4 g / litre (b) Metribuzin 3 g / litre
   (c) Both a & b (d) pendimethalin 3ml / lit
4. 2, 4-D sodium salt recommend to control
   (a) Aquatic weeds (b) *Parthenium*
   (c) Broad leaved weeds (d) All the above

III. Answer the following questions

1. List out the problematic weeds and write their management
2. Weed management practices for aquatic weeds
3. Weed management practices for parasitic weeds
4. Write type of parasitic weed with example
5. Integrated weed management practices for *Parthenium*

ANNEXURE I – MAJOR WEEDS IDENTIFICATION

GRASSES

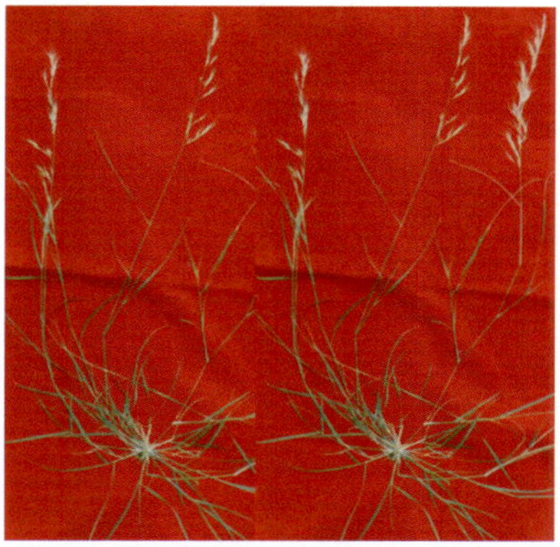

Aristida funiculata

Acrachne racemosa

Brachiaria reptans

***Chloris inflate: Syn:* Chloris barbata**

Cynodon dactylon

Dactyloctenium aegyptium

Digitaria bicornis

Digitaria longiflora

Dinebra arabica

Dinebra retroflexa

Echinochloa colona

Echinochloa crus-galli

Erigeron karvinskianus

Heteropogon contortus

Ischaemum indicum

Iseilema laxum

Panicum flavidum

Panicum javanicum

Panicum repens

Pennisetunm cenchroides

Perotis indica

Setaria verticiliata

Tragus roxburghii

SEDGES

Cyperus difformis

Cyperus iria

Cyperus nutans

Cyperus rotundus

Cyperus tenuculmis

Fimbristylis aestivalis

Fimbristylis argentea

Fimbristylis complanata

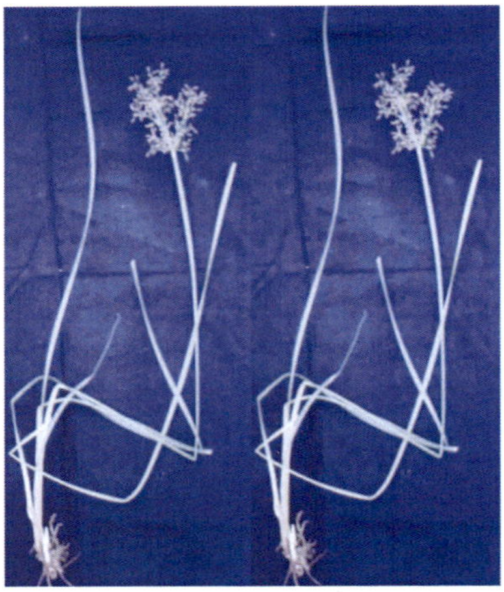

Fimbristylis milliacea

Fimbristylis woodrowii

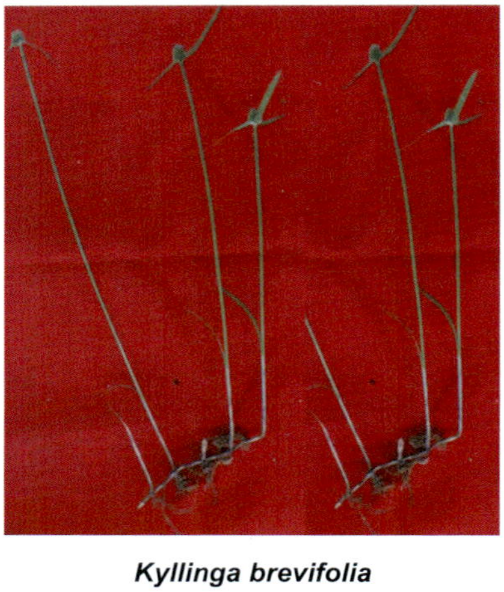

Kyllinga brevifolia

Kyllinga nemoralis

Pycerus polystachyos

Cyperus castaneus

BROAD LEAVED WEEDS

Abutilon hirtum

Abutilon indicum

Acalypha ciliata

Acalypha indica

Acanthospermum hispidum

Achyranthes aspera

Aerva lanata

Aerva tomentosa

Aeschynomene indica

Ageratum conyzoides

Alternanthera paronychioides

Alternanthera sessilis

Alternanthera pungens

Allmania nodiflora

Alysicarpus rugosus

Amaranthus spinosus

Amaranthus polygamus

Amaranthus viridis

Ammannia baccifera

Anisomeles malabarica

Andrographis paniculata

Andrographis serpyllifolia

Argemone mexicana

Aristolochia bractiata

Asystacia gangetica

Barleria montana

Bergia ammanioides

Bidens pilosa

Borreria hispida

Blepharis maderaspatensis

Blepharis molluginifolia

Boerhaavia diffusa

Boerhaavia erecta

Calotropis gigantea

Calotropis procera

Cardiospermum halicacabum Var. luridum

Cardiospermum halicacabum Var. microcarpum

Cassia auriculata

Cassia nigricans

Celosia argentia

Centella asiatica

Chrozophora rottleri

Cleome viscosa

Cleome gynandra

Cleome chelidonii

Clitoria ternatea

Coccinia grandis

Commelina benghalensis

Conyza ambigua

Convolvulus arvensis

Croton bonplandianum

Corchorus aestuans

Corchorus olitorius

Corchorus trilocularis

Corchorus tridens

Crotalaria pallida

Crotalaria verugosa

Cuscuta reflexa

Cynotis axillaris

Cynotis culculata

Cynotis cristata

D. metal

D. stramonium

Desmodium diffusum

Desmodium triflorum

Digera muricata Syn: D. arvensis

Ecbolium ligustrinum Syn: E.viride

Eclipta prostrata Syn: E. alba

Enicostema axillare

Erigeron asteroids

Erigeron arvinskianus

Euphorbia geniculata

Euphorbia hirta

Euphorbia hypercifolia

Euphorbia prostrata

Euphorbia pulcherima

Evolvulus alsinoides

Flaveria trinervia

Geniosporum tenuiflorum

Gisekia pharnaceoides

Gomphrena celosioides

Grangea maderaspatana

Heliotropium indicum

Hemidesmus indicus

Hybanthus enneaspermus

Hibiscus micranthus

Hibiscus vitifolius

Hygrophila auriculata

Hyptis suaveolens

Indigofera caerulea

Indigofera tirta

Indoneesiella echioides

Jatropha gossypiifolia

Ipomea alba

Ipomea carnea

Ipomea pes-trigridis

Ipomea sepiaria

Justicia prostrata

Justicia tranquebariensis

Lactuca runcinata

Lagasca mollis

Lantana camara

Leptadenia reticulata

Leucas aspera

Leucas urticaefolia

Leucas hirta

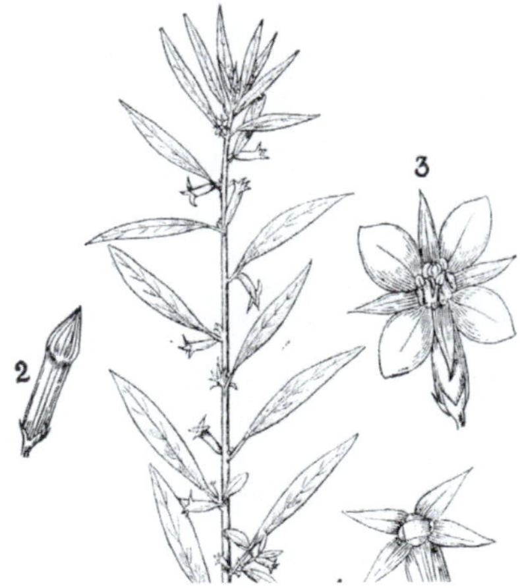

Ludwigia perennis

Malvastrum cromandlianum

Marselia quadrifolia

Melothria maderaspatnas

Merremia emarginata

Merremia tridenta

Mimosa pudica

Moniera cuneifolia

Mollugo disticha

Mollugo lotoides

Mollugo pentaphylla

Mollugo nudicaulis

Monochoria vaginalis

Moschosma polystachyum

Nastridium indicum

Ocimum canum

Ocimum tenuiflorum

Oldenlandia affinis

Oldenlandia alata

Oldenlandia umbellata

Orbanche cernua

Parthenium hysterophorus

Oxalis corniculata

Oxalis latifolia

Passiflora foetida

Pavonia zeylanica

Pedalium murex

Phyllanthus amarus

Phyllanthus maderaspatensis

Physalis minima

Phyla nodiflora

Portulaca oleracea

Portulaca quadrifida

Priva cordifolia

Rhynchosia minima

Rotala densiflora

Ruellia tuberosa

Rungia repens

Scoparia dulcis

Sida acuta

Sida cordifolia

Solanum eleaegnifolium

Solanum torvum

Solanum trilobatum

Solanum nigrum

Solanum xanthocarpum

Sonchus oleraceus

Sphaeranthus indicus

Spilanthes calva

Stachytarpheta jamaicensis

Striga asiatica

Stemodia viscosa

Synedrella nodiflora

Tephrosia purpurea

Tragia involucrata

Tragia plukenetii

Trianthema decandra

Trianthema protulacastrum

Trichodesma indicum

Tridax procumbens

Tribulus terrestris

Vernonia cinerea

Vicoa indica

Vigna trilobata

Waltheria indica

Wedelia chinensis

Xanthium strumarium